COMPUTATIONAL MODELING OF STRAND-BASED WOOD COMPOSITES IN BENDING

by

PEGGI LYNN CLOUSTON

B.A.Sc., The University of British Columbia, 1989
M.A.Sc., The University of British Columbia, 1995

A THESIS SUBMITTED IN PARTIAL FULFILMENT OF
THE REQUIREMENTS FOR THE DEGREE OF

DOCTOR OF PHILOSOPHY

in

THE FACULTY OF GRADUATE STUDIES
(Department of Forestry)

We accept this thesis as conforming to the required standard

THE UNIVERSITY OF BRITISH COLUMBIA

October 2001

ABSTRACT

A stochastic finite element approach is presented herein for simulating the nonlinear behaviour of strand-based wood composites with strands of varying grain angle. The approach is based on the constitutive properties of the individual strands. In this way, the approach is practical and versatile in application. It can be used to gauge the effects of varying strand characteristics (such as species or geometry) in product manufacturing or it may be useful in the design of wood composite structures (such as optimizing strand orientation for complex mechanical connections).

Within the thesis, both 2 and 3 dimensional, stochastic, materially nonlinear finite element codes are developed. The nonlinear constitutive behaviour of the wood strands is characterized within the framework of rate-independent theory of orthotropic elasto-plasticity. The constitutive model employs the Tsai-Wu yield criterion with the associated flow rule of plasticity. Upon plastic flow, anisotropic hardening is invoked whereby there is an expansion as well as distortion of the initial yield surface. Failure is marked by an upper bound surface whereupon either perfect plasticity (ie. ductile behaviour) or an abrupt loss of strength and stiffness (ie. brittle behaviour) ensues.

This constitutive model is implemented into the finite element codes. The programs are based on the conventional displacement formulation using linear isoparametric elements. The nonlinearities in the equilibrium equations are resolved by a modified Newton-Raphson procedure.

The programs are further formulated in a probabilistic manner using random variables as input. Principal material strength and stiffness properties (required in the yield criterion) are described by appropriate statistical distributions. To generate entire data samples for comparison with experimental samples, the programs were written with extended capacity to perform Monte Carlo simulations.

The mechanical properties of the strands, which are assumed to be orthotropic and homogeneous, are derived through both experiment and analysis. An experimental database of principal material strengths and stiffness parameters is acquired for Douglas-fir heartwood strands. Statistical parameters for shear strength and stiffness as well as the interaction parameter of the Tsai-Wu criterion are estimated, however, through a least square minimization of error between simulated and experimental compression strength of $[\pm 15]_s$ angle-ply laminates. Weibull weakest-link theory is employed to adjust experimental tensile strength values for size effect.

The general performance of the programs is verified through comparison of results for several analyses solved using analytical techniques or alternate programs. Following this, the ability of the models to reproduce experimental findings for angle-ply laminates in tension, compression and 3 point bending is validated. A preliminary investigation is conducted to compare numerical simulations with experimental data for Parallam® in tension and 3 point bending . The favourable comparisons of the model to experimental results attest to the effectiveness of the proposed technique.

TABLE OF CONTENTS

LIST OF FIGURES

LIST OF TABLES

NOTATION

The following symbols are used in this study:

[A] = direction cosine matrix;

a_i = defined in Equation 3.41;

a_{ij} = direction cosine components;

[B] = strain displacement matrix;

b = width of specimen;

b_i = randomly generated bivariate standard normal parameters;

[C] = 3 dimensional material stiffness matrix in principal material coordinates

[C'] = 3 dimensional material stiffness matrix in principal material coordinates

C_{ij} = components of 3 dimensional material stiffness tensor;

d = depth of specimen;

E_i = elastic moduli;

E_i' = tangent moduli;

F = load;

{F} = external force vector;

F_{ij} = components of strength tensor;

f = yield function;

G = in-plane shear modulus ;

g = plastic potential function;

H' = hardening modulus;

I = moment of Inertia;

[J] = Jacobian matrix;

[K] = global stiffness matrix of structure;

$[k]^e$ = element stiffness matrices;

k = threshold stress;

L = length;

L_{ij} = components of lower triangle of correlation matrix ;

M = bending moment;

M_{ij} = components of strength tensor;

m = scale parameter of Weibull distribution;

N_i = shape functions;

n = total number of layers;

P = number of probability levels;

{P} = internal resisting force vector;

[Q] = 2 dimensional material stiffness matrix in laminate coordinate system

[Q'] = 2 dimensional material stiffness matrix in principal material coordinate system

Q_{ij} = components of 2 dimensional material stiffness tensor;

{R} = resultant in-plane force vector;

R = ratio to proportion stress value;

S = in-plane shear strength;

SD = standard deviation;

[T] = transformation matrix;

t = thickness of lamina;

{u} = vector of nodal displacements;

u, v, w = displacement in x, y, z directions;

V = volume;

W_i = weighting factors for Gaussian integration;

W_{int}, W_{ext} = internal and external work;

W^p = plastic work;

X = parallel-to-grain strength;

X_t = parallel-to-grain tensile strength;

X_c, X_c^u = parallel-to-grain compressive yield and ultimate strength, respectively;

Y = perpendicular-to-grain strength;

Y_t = perpendicular-to-grain tensile strength;

Y_c, Y_c^u = perpendicular-to-grain compressive yield and ultimate strength, respectively;

z_i = standard normal random variable;

α_i = parameters which define the offset of the yield surface;

β = shape parameter of Weibull distribution;

γ_i = engineering shear strain components;

Δ = incremental value;

δ_y = vertical displacement;

ε_i = strain components;

ε_i^p = plastic strain components;

$\bar{\varepsilon}^p$ = effective plastic strain;

ζ = curvilinear coordinate of isoparametric element;

η = curvilinear coordinate of isoparametric element;

θ = grain angle;

Λ = vector of unknowns in minimization procedure;

λ = plastic multiplier;

μ = mean value;

υ_{ij} = Poisson's ratio;

ξ = curvilinear coordinate of isoparametric element;

Π = defined in Equation 3.53;

ρ_i = portion of effective stress;

σ_i = stress components;

$\bar{\sigma}$ = effective stress;

τ = non-specific material strength;

Φ = minimization function;

χ = hardening parameter;

$\{\Psi\}$ = residual force vector;

Subscripts and Superscripts

x, y, z = laminate coordinate directions;

1, 2, 3 = principal material directions;

c = designates center value;

e = designates elastic quantity;

ep = designates elastic-plastic quantity;

k = designates layer number;

max = designates maximum value;

o = designates original value;

p = designates plastic quantity;

r = iteration number;

u = designates ultimate value;

y = designates value at incipient yield.

ACKNOWLEDGEMENTS

I would like to extend a warm thank you to my supervisory committee, Dr. Frank Lam, Dr. Helmut Prion and Dr. Ricardo Foschi for their guidance and support throughout the project.

Thanks also go to members of my examining committee, Dr. Gary Schajer, Dr. J. David Barrett and Dr. Robert Leichti for reviewing my work and providing helpful commentary.

Acknowledgment is made to my friends and colleagues in the department of Wood Science, UBC, as well as our timber engineering lab manager, Bob Myronuk, who helped to make my graduate school days pleasantly memorable and rewarding.

My gratitude goes to my Mother for providing support and encouragement when most needed and for helping me stay focused and true to my goal throughout these years.

Finally, I wish to express my deepest gratitude to my best friend and husband, Alexander Schreyer, for his unwavering support, ever-insightful comments and always sage advice - both technical and personal.

Where order in variety we see
And where,
though all things differ,
all agree

Alexander Pope

1.1 INTRODUCTION

A new division of structural building materials is rapidly gaining recognition and acceptance in today's construction industry. Generically known as Structural Composite Lumber (SCL), this group consists of Laminated Veneer Lumber (LVL), Parallel Strand Lumber(PSL), Laminated Strand Lumber (LSL) and thick oriented strand board/rimboard. These new engineered products have important advantages over conventional lumber owing to their sophisticated manufacturing processes. They are typically made with thin wood veneers, strands or flakes which are arranged and bonded together using adhesive under controlled heat and pressure. Reconstitution of the wood in this manner disperses the natural defects throughout the material resulting in more consistent mechanical properties than that of conventional lumber. This high consistency leads to a more efficient utilization of the wood fiber resource. Consequently, structural composite lumber is replacing conventional lumber in traditional wood applications (*ie.* beams and columns) and, moreover, are starting to be used in applications typically dominated by steel or concrete (*ie.* long span commercial roof trusses and shell structures).

In contrast to the current growth of structural composite lumber, development of these products was relatively slow, depending predominantly on expensive empirical-based research initiatives. Typically, the products were refined gradually through experimental manipulation of the material constituents. The process was costly and time consuming. In light of this, an alternative approach for the development of future wood composite products is highly desired, if not outright necessary.

A logical alternative is the implementation of a computational model. Computer models can be used to estimate the effects of varying raw material characteristics on the final product's mechanical properties thereby reducing fabrication and testing costs for a new product. In the same manner, the model could be used to optimize or customize current products. For example, with an accurate working model, companies could gauge the effect of strategically placing lower quality or less expensive species in the board. Moreover, an accurate computer model could be an invaluable design aid for analyzing (or optimizing for) complex applications, such as connection details. The question then becomes one of developing an accurate working model.

Unfortunately, little work has been done in this area for wood-based composites. Only very recently have some studies attempted to analytically model relatively simple tensile specimens of strand or veneer-based composites. However, a vast amount of research has been done on the mechanical behaviour of advanced composite materials, (ie. those made from high modulus fibers such as graphite, silicon carbide, or boron, and used in aerospace applications). A multitude of modeling techniques and concepts have been established to characterize their stress-strain and failure behaviour, as discussed in any standard text on mechanics of composites (Jones, 1975; Hull, 1981; Gibson, 1993). One can fortunately draw ideas from this resource, while incorporating established wood mechanics principles.

Such is the intent of this thesis. Specifically, the focus of this thesis is the development of a viable model for predicting the mechanical response of a parallel-aligned wood strand composite under bending. This material is necessarily a simplified version, with convenient and controllable geometry, of the more complex material parallel strand lumber (PSL). Although beyond the scope of this thesis, the ultimate aim would be to provide a tool for commercial wood composites. As such, the current characteristics of the commercial product Parallam®, PSL have been used as a general reference from which to determine the model's primary considerations.

1.2   MODEL CONSIDERATIONS

Parallam®, PSL is a complex wood composite material fabricated through a technologically advanced processing technique. A visual depiction of the manufacturing process, taken from Sharp (1996), is shown in Figure 1.1. In forming a billet, the strands, ranging in length from approximately 0.6m to 2.4m, are delivered to a moving trough where they are randomly arranged through the depth and width and roughly aligned along the length. This deposition system results in a product with a highly complex geometric layout. From a macroscopic standpoint, the product could be considered as homogeneous and product behaviour characterized by averaged mechanical properties.

Figure 1.1 - Parallam®, PSL Manufacturing Process

However, this approach has no regard for the constituent materials (*ie.* the strands and glue). It lacks the ability to account for varying properties or geometric arrangement of the strands in order to customize composite performance. Ergo, the present model's methodology is based on the mechanical properties of the individual strands and the geometric lay up is implicit in the analysis. The strand properties are assumed to be homogeneous within each strand and are established empirically. It is noted that although gross longitudinal misalignment of strands is rare in PSL, gross grain deviation *is* considered in this thesis because individual strands may contain them, and future product development may investigate variation of fibre orientation to improve transverse strengths and stiffnesses.

The strand deposition system also results in a product with high strength and stiffness in the longitudinal direction and comparatively poor properties in other directions. Macroscopically, wood is assumed to be orthotropic whereby properties differ in three orthogonal directions only, (Bodig and Jayne, 1982; Perkins, 1967) which greatly simplifies the stress-strain relationships for the material. Similarly, it is assumed that PSL is orthotropic with principal material coordinates as shown in Figure 1.2.

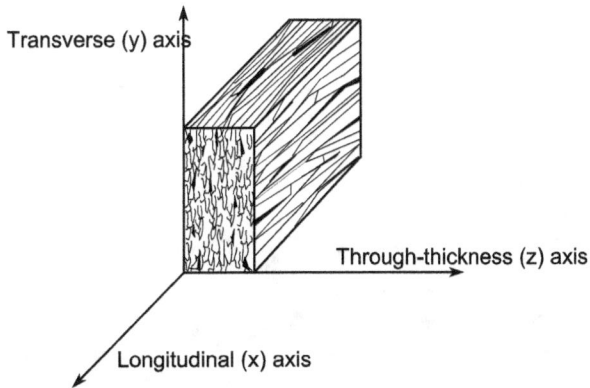

Figure 1.2 - Principal Material Coordinates of PSL

The mechanical properties of the strands are inherently variable. This variability can significantly influence the accuracy of the strength prediction model and should be included integrally. In this thesis, parameter variation is handled through a probabilistic approach whereby measured quantities are expressed in statistical terms, known as 'random variables'. Probabilistic solutions are obtained through Monte Carlo simulations.

When using data obtained from small specimens to predict the strength of large members, one must consider the variation of strength with size. This 'size effect' is most prominent with brittle modes of failure and is dependent on the strength variability of the material. Size effect has been acknowledged and addressed in many wood and timber studies such as, Barrett et al. (1975), Madsen and Buchanan (1986), and Barrett et al. (1995), as well as studies on structural composite lumber by Sharp and Suddarth (1991), and Clouston et al. (1998). Each of these studies espoused the use of Weibull weakest link theory to quantify size effect. As such, this theory is adopted for use in the current study.

1.3   OBJECTIVE AND SCOPE

The fundamental objective of the thesis is to develop a computational model for simulating the bending behaviour of parallel-aligned wood strand composites. An orthotropic finite element code will be developed for the general nonlinear analysis of wood laminates with layers of varying fiber orientation. The code will be developed within a probabilistic framework using random variables as input enabling stochastic-based Monte Carlo simulation analyses. It will be shown that orthotropic plasticity theory is effective in describing the nonlinear behaviour up to and beyond the point of laminate failure.

Chapter 2 will provide a general review from the literature on the background of the yield criterion chosen for use in the constitutive model, as well as an outline of published studies on modeling of wood composites. Chapter 3 aims at a complete description of the elastic-plastic-failure constitutive model employed in the model. Chapter 4 provides a review of the finite element formulation as well as a thorough program performance verification. Chapter 5 outlines the acquisition of the strand database while chapter 6 details application of the model to a series of problems. The numerical simulations are compared to experimental data for validation. Finally, in chapter 7, a summary is provided as well as a discussion on further areas of research.

2.1 INTRODUCTION

Despite its importance, only a small number of studies have focused on strength prediction of structural wood composites. These few studies, all the same, employed conventional strength analysis techniques which have been proven successful for laminated composite materials. The basic building block in the analysis is the mechanical behavior of the individual layer, or lamina, (Figure 2.1) which can be studied from either a micro-mechanical or macro-mechanical perspective.

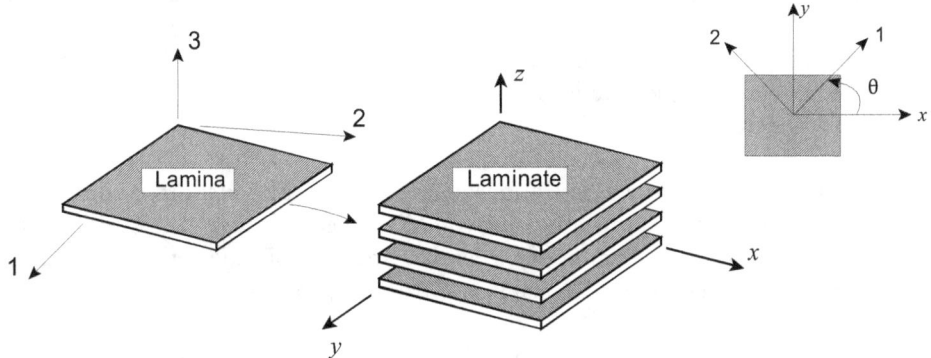

Figure 2.1 - Lamina and Laminate Coordinate System

Micro-mechanical theories determine the mechanical properties of a lamina by addressing constitutive information about the fiber and the matrix phase. The interaction of the phases is examined in detail through mechanics of materials principles - for example, the rule of mixtures - to determine strength and stiffness properties of the heterogeneous material.

Macro-mechanical theories, on the other hand, involve the use of gross effective lamina properties wherein the material is presumed homogeneous. The effects of the constituent materials as well as inherent cracks, notches or other discontinuities which lead to failure are not directly modeled; rather, they are incorporated indirectly in an averaged or "smeared" manner.

A macro-mechanical approach has been adopted for this thesis. Lamina properties are established empirically and used in a semiempirical failure criterion. This chapter presents first an introduction to orthotropic failure criteria, which are integral in a macro-mechanical analysis, followed by a review of the specific criterion used for the present model, and finally a survey of wood composite publications focused on material behaviour modeling.

2.2 FAILURE THEORIES

Failure theories are, in general, mathematical models which may be determined theoretically by rational modeling of the material's physical characteristics or empirically by simply representing experimental observations. In the latter case, the theories are termed phenomenological 'treating the heterogeneous material as a continuum' and making no attempt to explain the mechanisms which lead to material failure (Wu, 1974).

Early classical theories were based on limiting a physical variable such as stress, strain, or strain energy assuming a homogeneous and isotropic material. For example, one of the first theories, the Maximum Normal Stress theory, maintains that material failure occurs when the maximum normal stress reaches a critical value independent of other stresses at that point. The critical value can be determined from a uniaxial tensile test. Three simple, independent equations result:

$$\sigma_1 = \sigma_c \ ; \ \sigma_2 = \sigma_c \ ; \ \sigma_3 = \sigma_c \tag{2.1}$$

where; σ_i (i = 1, 2, 3) denote principal stresses and σ_c is the critical stress. These equations may be plotted with each normal stress component constituting an axis forming a triad in the stress space. This is known as a failure envelope (or surface). Combinations of stresses contained inside the failure envelope signify survival and on or beyond the envelope indicate material failure. The concept is common for both isotropic and orthotropic criteria, however, for the latter the plotted stresses usually correspond to those along the principal material axes.

Many orthotropic criteria are based on isotropic criteria. The first orthotropic theory, devised by Hill in 1948, extended the well known isotropic von Mises theory for use with specially orthotropic materials (ie. materials for which the direction of the applied stresses coincide with the principal material directions). Hill formulated the criterion such that if the orthotropy vanished, the theory reduced to von Mises theory.

The von Mises theory was also used by C.B. Norris in 1950 to develop a strength criterion for orthotropic materials.[1] His theory applied the isotropic failure criterion to a simplified geometrical model. By his own admission, Norris's approach was not 'rigorously correct', representing an orthotropic material by an isotropic material with regularly spaced voids. Nonetheless, his equation produced good results when compared to data for 3 ply plywood and fiberglass laminate for which most of the glass fibers were aligned in one direction (Norris, 1962). Norris's theory has been recommended for design of glued-laminated beams in the 1974 Timber Construction Manual. However, it was also criticized for being overly conservative for many beam and arch situations in practice (Kobetz and Krueger, 1976).

There are numerous orthotropic yield theories available in the literature as compiled and compared in surveys by researchers Sandhu (1972), Rowlands (1985) and Nahas (1986). Each criterion has strengths and limitations and no one criterion is suitable for all materials. However, one criterion, made popular by Tsai and Wu in 1971, received widespread attention due to its simplicity and generality. The two-dimensional (i.e. plane stress) form of this theory will be used in this study.

2.2.1 Tsai-Wu criterion

Tsai and Wu proposed a simplified version of a tensor polynomial. Their theory predicts that failure will occur when the following inequality is satisfied:

$$F_i \sigma_i + F_{ij} \sigma_i \sigma_j \geq 1 \tag{2.2}$$

where; i, j = 1, 2, ... 6 (repeated indices imply summation), and F_i and F_{ij} are second and fourth rank strength tensors, respectively. Under plane stress, σ_3, σ_4, and σ_5 are assumed negligible and the F_6, F_{16} and F_{26} terms are zero due to special orthotropy. In expanded form Eq. (2.2) becomes:

---

[1]This theory succeeded his so called 'interaction formula' suggested in 1945 for use with plywood plates.

$$F_1\sigma_1 + F_2\sigma_2 + F_{11}\sigma_1^2 + F_{22}\sigma_2^2 + 2F_{12}\sigma_1\sigma_2 + F_{66}\sigma_6^2 \geq 1 \qquad (2.3)$$

The coefficients F_1 through F_{66}, with the exception of F_{12}, are described in terms of the strengths in the principal material directions. For a lamina in plane stress, these are: longitudinal strength in both tension and compression (X_T, X_C), transverse strength in both tension and compression (Y_T, Y_C) and in-plane shear strength (S). Considering a uniaxial tension load on a specimen in the 1 direction, the above equation at failure becomes:

$$F_1 X_T + F_{11} X_T^2 = 1 \qquad (2.4)$$

and for compression is:

$$F_1 X_C + F_{11} X_C^2 = 1 \qquad (2.5)$$

By solving the equations 2.4 and 2.5 simultaneously and regarding the compression strength as negative, the expression for the strength parameters F_1 and F_{11} are found to be:

$$F_1 = \frac{1}{X_T} - \frac{1}{X_C} \; ; \; F_{11} = \frac{1}{X_T X_C} \qquad (2.6)$$

Through similar mathematical manipulations, it can be shown that:

$$F_2 = \frac{1}{Y_T} - \frac{1}{Y_C} \; ; \; F_{22} = \frac{1}{Y_T Y_C} \; ; \; F_{66} = \frac{1}{S^2} \qquad (2.7)$$

A drawback in the application of the tensor polynomial theory is that there is no consensus among researchers for a method of determining the value of the combined stress parameter, F_{12} which accounts for the interaction between normal stresses, σ_1 and σ_2. The only certainty is that in order for the failure surface to be a closed ellipsoid[2], the coefficient F_{12} must be bounded by the stability condition:

$$F_{11}F_{22} - F_{12}^2 \geq 0 \qquad (2.8)$$

The 'correct' determination of F_{12} has been the topic of discussion for many researchers for many years.

2.2.1.1   Significance of the Interaction Term, F_{12}

The interaction term, F_{12} characterizes the interaction of the normal stresses, σ_1 and σ_2. Because this term involves both normal stresses, its evaluation must occur under a biaxial loading situation unlike the other strength tensors whose values are determined from uniaxial loading. The Tsai-Wu failure envelope takes the form of an ellipse in the σ_1, σ_2 plane as shown in Figure 2.2. The strength tensors F_1, F_2, F_{11} and F_{22} establish the axes intercepts whereas the component F_{12} determines the rotation of the ellipsoid with respect to the principal material axes. Figure 2.2 shows the influence of varying F_{12} in equation 2.2 when

---

2. Plotting equation 2.3 with σ_1 and σ_2 as coordinates of a point will produce an ellipsoidal failure surface, providing F_{12} is within the prescribed range. Else, the surface would become open-ended, meaning no matter how large the stresses became, failure would never ensue, which is physically impossible.

no shear stress is present. It has been argued that the value of F_{12} "determines the effectiveness of tensorial-type failure criteria." (Suhling et al., 1984)

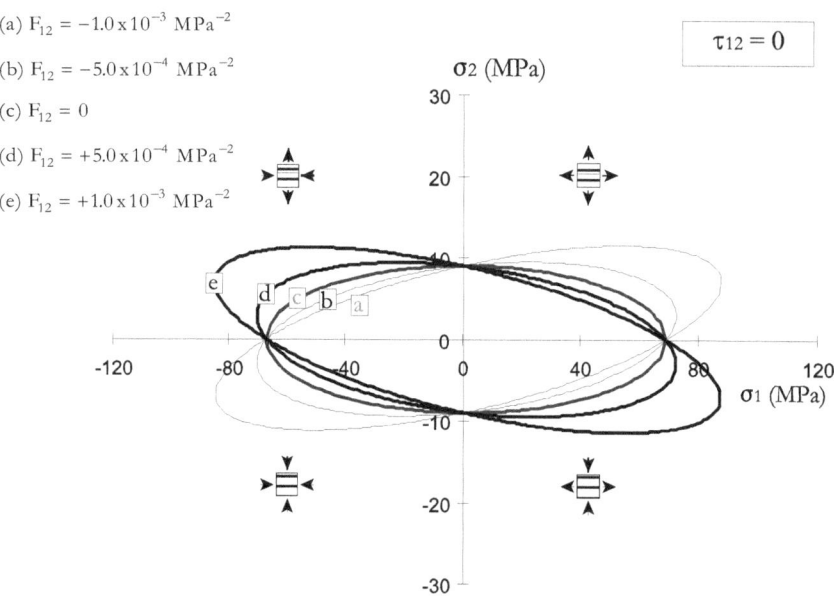

(a) $F_{12} = -1.0 \times 10^{-3}\ MPa^{-2}$

(b) $F_{12} = -5.0 \times 10^{-4}\ MPa^{-2}$

(c) $F_{12} = 0$

(d) $F_{12} = +5.0 \times 10^{-4}\ MPa^{-2}$

(e) $F_{12} = +1.0 \times 10^{-3}\ MPa^{-2}$

Figure 2.2 - Theoretical Strength Envelopes for Wood with Different F_{12} Values

2.2.1.2   Evaluation of interaction term, F_{12}

Tsai and Wu (1971) reported that a separate combined stress test was necessary for determining F_{12}. They provided six viable test options in their original paper: a simple biaxial tension or compression test with the applied load equal in each direction; a 45 degree off-axis tension or compression test; or a 45 degree off-axis positive or negative shear test as illustrated in Figure 2.3.

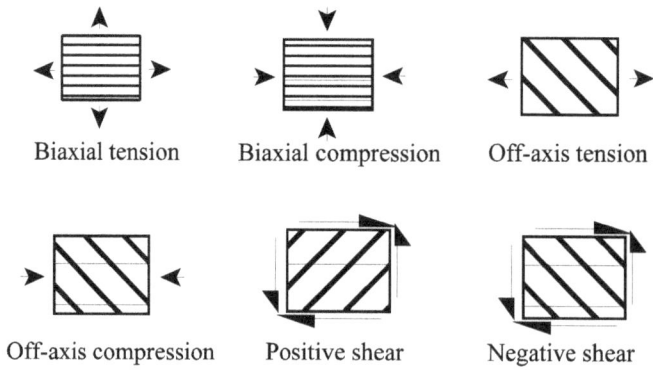

Figure 2.3 - Six Original Tests Proposed by Tsai and Wu (1971)

Strict care must be taken in determining F_{12} due to its high sensitivity to experimental variation. Tsai and Wu showed that for a graphite epoxy composite, slight inaccuracies in measurements of strength for most of the above tests would result in large inaccuracies in the calculated value of F_{12}. This problem is accentuated by the fact that in order for F_{12} to be physically admissible, it must satisfy the prescribed stability bounds (Eqn. 2.8). Many researchers conducted similar studies on a variety of materials in an attempt to determine F_{12} experimentally.

Pipes and Cole (1973) performed a limited number of off-axis tensile tests on boron epoxy composites to determine F_{12}. They tested two coupons at 60 degrees, three at 45 degrees, four at 30 degrees, and three at 15 degrees. Using the mean of each set they found that only the value obtained for the 15 degree off-axis tests satisfied the stability criterion. Consequently, they concluded that the off-axis tensile test was not an adequate method of determining F_{12} for boron epoxy composites.

Suhling, et al. (1984) conducted a comprehensive study to establish F_{12} for paperboard employing all of the six test methods suggested by Tsai and Wu (1971). They found the optimum value of F_{12} by fitting a linear regression through experimental results for all four quadrants and compared this value with that obtained from the individual tests. The tension biaxial tests produced the closest value to the optimum value for a zero shear situation. They found that $F_{12} = 0$ was a satisfactory solution for four shear levels considered; $\sigma_6 = 0$, 6.9, 10.3, and 15.9 MPa. They also concluded that due to the highly sensitive and unstable nature of F_{12} when calculated using off-axis tests that this test was not a suitable method to determine the interaction parameter.

Anticipating the problems that researchers might encounter when determining F_{12}, Wu published a paper in 1974 demonstrating a method to find an optimal biaxial stress ratio for calculating F_{12} experimentally. This method was not strongly espoused because it involved a series of experimental iterations further complicating the tensor polynomial theory. Difficulties in experimental determination of F_{12} prompted several researchers to seek a theoretical solution to the problem.

Narayanaswami and Adelman (1977) performed a numerical analysis to prove that, from a practical standpoint, the arbitrary assignment $F_{12} = 0$ was acceptable for filamentary composites. For the six tests described above, they computed the percentage error when setting $F_{12} = 0$ for 10 different composite materials. In all cases the error was found to be less than ten percent and they therefore concluded this to be an acceptable error for practical engineering applications.

Cowin (1979) derived Hankinson's formula:

$$\sigma_\theta = \frac{XY}{X \sin^2 \theta + Y \cos^2 \theta} \qquad (2.9)$$

(an empirical equation which has had remarkable and repeated success in predicting wood strength, σ_θ, at an angle to grain, θ) using only the linear term of the tensor polynomial function, $F_i \sigma_i$. By considering the quadratic terms as well, he derived a formula for F_{12}:

$$F_{12} = \sqrt{F_{11} F_{22}} - \frac{1}{2S^2} \qquad (2.10)$$

that produced a similar, yet more accurate, equation to Hankinson's criterion to represent bone strength with respect to angle to grain.

Both van der Put (1982) and Liu (1984) derived an expression for F_{12} that reduces the tensor polynomial theory into the Hankinson formula:

$$F_{12} = \frac{1}{2}\left(\frac{1}{X_T Y_C} + \frac{1}{X_C Y_T} - \frac{1}{S^2}\right)$$

(2.11)

van der Put approached the problem by transforming the applied stresses to the principal material coordinate system while Liu transformed the strength tensors.

Despite all efforts, no standard method to determine F_{12} has been established. As such, a probabilistic approach has been developed in this study and will be addressed in chapter 5.

2.3   LITERATURE REVIEW OF WOOD COMPOSITE MODELING

One of the earliest attempts to analytically predict wood composite behaviour was by Hunt and Suddarth (1974). They predicted tensile modulus of elasticity and shear modulus of rigidity of medium density homogeneous flakeboard. The method involved a linear elastic finite element analysis of a tensile specimen modeled as an assemblage of four noded plate elements (representing the flakes) embedded in a rectangular grid of rigidly connected beam elements (representing the resin). The grain orientation in each plate element was randomly assigned simulating a random flake deposition in the mat. Young's modulus was calculated using Hooke's law where the stresses are simply the applied stress over the area and strain is the analog's average axial strain, Poisson's ratio was calculated from the usual relation $v = \epsilon_1 / \epsilon_2$ and shear modulus was determined from the isotropic relation $G = E / 2(1+v)$. A Monte Carlo technique was used to produce distributions of the desired values (E and G) and the average analytical values were compared with experimentally obtained averages for two separate species, Douglas-fir and aspen. The model underestimated the experimental tensile modulus by 8 percent for aspen and 6 percent for Douglas-fir whereas the shear modulus was overestimated by 10 percent and 13 percent for aspen and Douglas-fir respectively.

More recently, Shaler and Blankenhorn (1990) evaluated the suitability of two composite theories - the rule of mixtures and the Halpin-Tsai equation - to predict flexural modulus of elasticity of oriented flakeboard. They considerd flake geometry and orientation, density, resin content and species. The modeling approach was a two-stage process. The first stage treated the resin as matrix and the wood as fiber and the second stage treated the resin and wood as matrix and air as fiber. The method underestimated measured moduli by an average of 25 percent.

In 1993, Triche and Hunt developed a linear elastic finite element model capable of predicting the tensile strength and stiffness of a PWSC with controlled geometry. The model was micromechanical in nature considering each strand to have three layers (ie. pure wood, resin and an interface between them) and used, as input, the properties of the individual constituents. For efficiency, the concept of a 'superelement', (an element representing one entire strand system) was introduced. Superelements were built by forming a mesh of elements for the model of one strand and "statically condensing internal degrees of freedom so that the remaining model (contained) only degrees of freedom that (were) essential in building the board model." Composite boards were constructed from a group of superelements and corresponding board stiffness (using the stiffness matrices of the superelements) was computed in a conventional finite element procedure.

A large experimental database on the longitudinal tensile stiffness (and to a lesser extent, strength) of yellow poplar strands was developed. Also, a smaller sample of laminated assemblies was developed to assess the thickness change due to pressing of the strands and to acquire the properties of the wood-resin interface. The model was enhanced to consider the effect of overlapping strands. This was handled by "building on the existing model until its predicted E matched the ..." mean value for a sample of lap sandwich specimens of the average E within the specimen.

The veracity of the model was checked through comparison of predicted and experimental results of 14 distinct categories of small composite boards with controlled geometry. Maximum strength was predicted according to several failure criteria including maximum stress theory and Tsai-Wu theory and board elasticity was calculated based on the nodal displacements of the model. Excellent accuracy was reported for the predicted E (from 0.0 to 11.1% error); however, prediction of maximum stress was inconsistent and, in at least one case, unacceptable (from 1.2 to 101.1% error).

Cha and Pearson (1994), developed a two-dimensional finite-element model which could predict the elastic tensile properties of a 3 ply veneer laminate consisting of an off-axis core ply of varying angles with or without a crack and clear, straight-grained outer plys. The laminate was modeled as a single plate and the stress components were defined across the laminate thickness. Good agreement was obtained between predicted and experimental strains at maximum load (maximum difference of 14.3%) as well as predicted and experimental stresses (maximum difference of 7.7%). Further, the model was used to investigate the influence of cracks and grain angle of the middle ply. For this, stresses within the layers were estimated as the product of the effective strain from the model and the corresponding elastic modulus of the lamina, determined experimentally. Resulting transverse and shear stresses were charted for varying grain angles and specified crack lengths.

Recently, a very progressive study was carried out by Wang and Lam (1998) in which a three dimensional, nonlinear, stochastic finite element model to estimate the probabilistic distribution of PWSC tensile strength was developed. The model is based fundamentally upon longitudinal tensile strength and stiffness data of individual strands. A single strength modification factor (α) was incorporated, defined as a function of individual strength factors (α_L, α_T, $\alpha_{int.}$, α_{oth}) as well as the number of plys in the composite. The individual strength factors accounted for, respectively, 1) Length and 2) Thickness effect (defined by Weibull weakest-link theory), 3) the presence of resin and its influence at the interface, and 4) the effect of other factors, estimated by way of a finite element analysis together with nonlinear, least square regression with experimental data for two and three ply assemblies.

The finite element model considered strictly longitudinally laminated strands employing a three dimensional layered element with 8 nodes. Load was applied incrementally and a simple, uniaxial failure criterion (a function of the strength modification factor) was used to establish individual element failure. Upon detection of failure, a nonlinear algorithm was implemented to calculate the subsequent displacements and hence strains and stresses depending on the modified structural stiffness matrix.

The model was verified through comparison with experimental data for four and six ply laminates. Further, various finite element grids were used, (affecting the size factor) to ensure the robustness of the prediction model. It was reported that, in all cases, excellent agreement was found.

3.1 INTRODUCTION

It is of interest to note that in the past, research and research guidelines (e.g. ASTM D143 and D198) for wood and timber bending strength have focussed to a large degree on linear elastic constitutive theory. This is logical in that, as first indicated by Perkins (1967), the bending strength of the conventional building material, timber, is generally governed by strength reducing flaws such as knots or slope of grain contributing to a brittle, tensile failure, which is accepted as being linear elastic (Madsen and Buchanan 1986; Barrett et al. 1995). However, the stress-strain behaviour of wood in compression, both parallel and perpendicular to grain, is known to be nonlinear (Goodman and Bodig, 1971; Maghsood et. al., 1973; Conners, 1988). Consequently, bending fracture in high grade timber (with few defects) is considered to be quasi-brittle, where the load-displacement behaviour may be nonlinear but ultimate failure is brought on by brittle fracture.

It is speculated that due to the minimized and controlled dispersement of flaws in wood composites such as PSL, compressive stresses play a more prominent role in a bending analysis when compared to that for timber. Hence, restriction of a bending analysis for a wood composite to the elastic range would result in an inefficient design because it would not account for the material strength beyond the proportional limit.

In light of this, the constitutive model for the wood laminae of this study is comprised of four basic behavioural domains: elastic, elasto-plastic, post-failure brittle, or post-failure ductile, as illustrated by a simple uniaxial stress-strain curve in Figure 3.1.

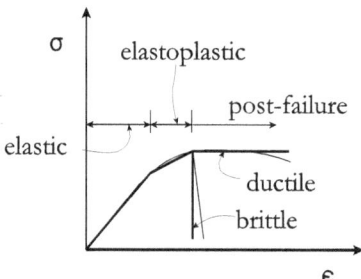

Figure 3.1 - Constitutive Model for Wood Laminae

It is presumed therein, that damage which occurs beyond the proportional limit can be described within the framework of classical plasticity theory. Upon initial loading, the material is assumed to behave linearly-elastic. Beyond the elastic domain the material may either 1) fail in a brittle manner or 2) first strain-harden and then ultimately fail in either a ductile or a brittle mode. For ductile failure, the material is assumed to lose all stiffness but retains strength, whereas for brittle failure both stiffness and strength are lost. Strain-hardening is characterized by successive growth of the yield surface defined by the Tsai-Wu criterion. The objective of this chapter is to derive the constitutive relationships for each behavioural regime.

3.2   CONSTITUTIVE MODEL

3.2.1   Linear elastic regime

The fundamental constitutive equation to describe linear elastic material behaviour is the well known Hooke's law, written here in generalized tensorial form (Sokolnikoff, 1956),

$$\sigma_{ij} = C_{ijkl}\varepsilon_{kl} \; ; \quad i,j,k,l = 1,2,3 \tag{3.1}$$

where repeated indices imply summation. This law defines a one-to-one mathematical relationship between the nine components of the stress tensor, σ_{ij} and the nine components of the strain tensor, ϵ_{ij}. The stiffness tensor, C_{ijkl}, is of 4th order, consisting of 81 stiffness coefficients.

From equilibrium requirements, both stresses and strains are symmetric (i.e., $\sigma_{ij} = \sigma_{ji}$ and $\epsilon_{ij} = \epsilon_{ji}$), so that there are truly only six independent components of each. This means that the stiffness tensor is also symmetric with respect to these indices (i.e. $C_{ijkl} = C_{jikl} = C_{ijlk} = C_{jilk}$) resulting in 36 stiffness coefficients. For the convenience of general mathematical manipulation, the double indexed system is often replaced by a single indexed system having a range of i=1,2,....6 as follows:

$$
\begin{aligned}
\sigma_{11} &= \sigma_1 & \varepsilon_{11} &= \varepsilon_1 \\
\sigma_{22} &= \sigma_2 & \varepsilon_{22} &= \varepsilon_2 \\
\sigma_{33} &= \sigma_3 & \varepsilon_{33} &= \varepsilon_3 \\
\sigma_{12} &= \sigma_4 & 2\varepsilon_{12} = \gamma_{12} &= \varepsilon_4 \\
\sigma_{13} &= \sigma_5 & 2\varepsilon_{13} = \gamma_{13} &= \varepsilon_5 \\
\sigma_{23} &= \sigma_6 & 2\varepsilon_{23} = \gamma_{23} &= \varepsilon_6
\end{aligned}
\tag{3.2}
$$

where γ_{23}, γ_{13}, and γ_{12} denote standard engineering shear strains, which physically represent the distortional change in angle from 90°. Equation (3.1) can now be written in contracted notation:

$$\sigma_i = C_{ij}\varepsilon_j \; ; \quad i,j = 1,2,...6 \tag{3.3}$$

It can be shown that the total possible number of independent stiffness coefficients may be reduced further from 36 to 21, independent of material symmetry. Given the strain energy density function, W, such that the stresses are defined as:

$$\sigma_i = \frac{\partial W}{\partial \varepsilon_i} = C_{ij}\varepsilon_j \tag{3.4}$$

we differentiate with respect to ε_j, to find

$$\frac{\partial^2 W}{\partial \varepsilon_i \partial \varepsilon_j} = C_{ij} \tag{3.5}$$

If we reverse the order of differentiation, we have

$$\frac{\partial^2 W}{\partial \varepsilon_j \partial \varepsilon_i} = C_{ji} \tag{3.6}$$

As the order of differentiation on W has no influence on the result, $C_{ij}=C_{ji}$ and the stiffness matrix is symmetric, reducing the total number of independent coefficients to 21.

Wood, in general, is assumed to have three orthogonal planes of symmetry. When the stress-strain relationships are developed within this principal coordinate system, there is no interaction between normal stresses and shearing strains, nor between shearing stresses and normal strains or shearing strains. As such, the number of independent components of the stiffness tensor reduces from 21 to 9 as follows:

$$\begin{Bmatrix} \sigma_1 \\ \sigma_2 \\ \sigma_3 \\ \sigma_4 \\ \sigma_5 \\ \sigma_6 \end{Bmatrix} = \begin{bmatrix} C_{11} & C_{12} & C_{13} & 0 & 0 & 0 \\ C_{12} & C_{22} & C_{23} & 0 & 0 & 0 \\ C_{13} & C_{23} & C_{33} & 0 & 0 & 0 \\ 0 & 0 & 0 & 2C_{44} & 0 & 0 \\ 0 & 0 & 0 & 0 & 2C_{55} & 0 \\ 0 & 0 & 0 & 0 & 0 & 2C_{66} \end{bmatrix} \begin{Bmatrix} \varepsilon_1 \\ \varepsilon_2 \\ \varepsilon_3 \\ \gamma_{12}/2 \\ \gamma_{13}/2 \\ \gamma_{23}/2 \end{Bmatrix} \tag{3.7}$$

The stiffness coefficients, in terms of engineering constants (i.e. Young's moduli (E), shear moduli (G), and Poisson's ratio (v)) are:

$$C_{11} = \frac{1 - \upsilon_{23}\upsilon_{32}}{E_2 E_3 \Delta} \qquad C_{12} = \frac{\upsilon_{21} + \upsilon_{31}\upsilon_{23}}{E_2 E_3 \Delta} \qquad C_{13} = \frac{\upsilon_{31} + \upsilon_{21}\upsilon_{32}}{E_2 E_3 \Delta}$$

$$C_{22} = \frac{1 - \upsilon_{13}\upsilon_{31}}{E_1 E_3 \Delta} \qquad C_{23} = \frac{\upsilon_{32} + \upsilon_{12}\upsilon_{31}}{E_1 E_3 \Delta} \qquad C_{33} = \frac{1 - \upsilon_{12}\upsilon_{21}}{E_1 E_2 \Delta} \tag{3.8}$$

$$C_{44} = G_{12} \qquad C_{55} = G_{13} \qquad C_{66} = G_{23}$$

where

$$\Delta = \frac{1 - \upsilon_{12}\upsilon_{21} - \upsilon_{23}\upsilon_{32} - \upsilon_{31}\upsilon_{13} - 2\upsilon_{21}\upsilon_{32}\upsilon_{13}}{E_1 E_2 E_3} \tag{3.9}$$

Poisson's ratios are dependent and can be estimated through the reciprocal relationship

$$\frac{\upsilon_{ij}}{E_i} = \frac{\upsilon_{ji}}{E_j} \qquad i, j = 1, 2, 3 \tag{3.10}$$

Lamina analysis often assumes a two-dimensional stress-state (plane stress) where σ_3, σ_5, and $\sigma_6 = 0$. Considering this, the lamina stiffness matrix, denoted Q_{ij}, contains only four independent components as follows:

$$\left\{\begin{array}{c} \sigma_1 \\ \sigma_2 \\ \sigma_4 \end{array}\right\} = \begin{bmatrix} \dfrac{E_1}{1-\upsilon_{12}\upsilon_{21}} & \dfrac{\upsilon_{12}E_2}{1-\upsilon_{12}\upsilon_{21}} & 0 \\ \dfrac{\upsilon_{12}E_2}{1-\upsilon_{12}\upsilon_{21}} & \dfrac{E_2}{1-\upsilon_{12}\upsilon_{21}} & 0 \\ 0 & 0 & 2G_{12} \end{bmatrix} \left\{\begin{array}{c} \varepsilon_1 \\ \varepsilon_2 \\ \gamma_{12}/2 \end{array}\right\} = [Q] \left\{\begin{array}{c} \varepsilon_1 \\ \varepsilon_2 \\ \gamma_{12}/2 \end{array}\right\} \qquad (3.11)$$

3.2.1.1 Constitutive Relations for Off-axis Coordinates

If the stresses and strains are defined in some non-principal material directions, (or off-axis coordinates) such as x and y in Figure 2.1, as is often the case for a lamina within a laminated member, the lamina is termed 'generally' orthotropic. The stiffness matrix in this case is fully populated (ie. no zero terms present) and can be derived in terms of C_{ij} by considering the transformation laws for Cartesian tensors as follows:

If we consider two rectangular coordinate systems with a common origin as shown in Figure 3.2, we can calculate the relative orientation of one axis with respect to the other through use of the direction cosines a_{ij}=cos (x_i, x_j'). In the case of the lamina in Figure 3.2,

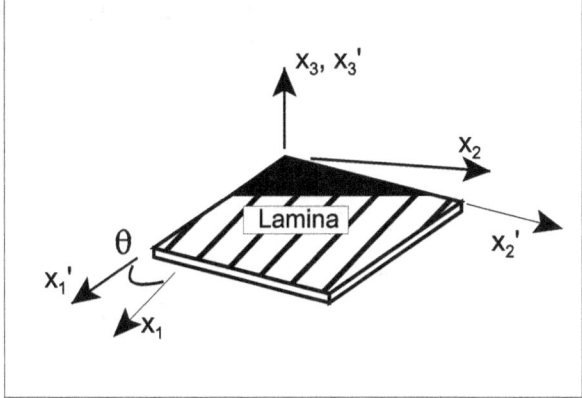

Figure 3.2 - **Principal and Off-axis Coordinate Systems of a Lamina**

$$a_{ij} = A = \begin{bmatrix} a_{11} & a_{12} & a_{13} \\ a_{21} & a_{22} & a_{23} \\ a_{31} & a_{32} & a_{33} \end{bmatrix} = \begin{bmatrix} \cos\theta & \sin\theta & 0 \\ -\sin\theta & \cos\theta & 0 \\ 0 & 0 & 1 \end{bmatrix} \qquad (3.12)$$

Referencing Figure 2.1, we can now transform the stress (or strain) tensor from the laminate axes (x,y,z) to the principal axes (1,2,3), using the direction cosine matrix in accordance with the transformation law (Mase, 1970),

$$\begin{bmatrix} \sigma_{11} & \sigma_{12} & \sigma_{13} \\ \sigma_{12} & \sigma_{22} & \sigma_{23} \\ \sigma_{13} & \sigma_{23} & \sigma_{33} \end{bmatrix} = [A] \begin{bmatrix} \sigma_x & \sigma_{xy} & \sigma_{xz} \\ \sigma_{xy} & \sigma_y & \sigma_{yz} \\ \sigma_{xz} & \sigma_{yz} & \sigma_z \end{bmatrix} [A]^T \qquad (3.13)$$

Simplifying the right hand side and using contracted notation, we have

$$
\begin{Bmatrix} \sigma_1 \\ \sigma_2 \\ \sigma_3 \\ \sigma_4 \\ \sigma_5 \\ \sigma_6 \end{Bmatrix} = \begin{bmatrix} c^2 & s^2 & 0 & 2sc & 0 & 0 \\ s^2 & c^2 & 0 & -2sc & 0 & 0 \\ 0 & 0 & 1 & 0 & 0 & 0 \\ -sc & sc & 0 & c^2-s^2 & 0 & 0 \\ 0 & 0 & 0 & 0 & c & s \\ 0 & 0 & 0 & 0 & -s & c \end{bmatrix} \begin{Bmatrix} \sigma_x \\ \sigma_y \\ \sigma_z \\ \sigma_{xy} \\ \sigma_{xz} \\ \sigma_{yz} \end{Bmatrix} = [T]\{\sigma'\}
\tag{3.14}
$$

where $c \equiv \cos\theta$ and $s \equiv \sin\theta$ and [T] is termed the transformation matrix. The stresses in the x-y system can be written

$$
\begin{Bmatrix} \sigma_x \\ \sigma_y \\ \sigma_z \\ \sigma_{xy} \\ \sigma_{xz} \\ \sigma_{yz} \end{Bmatrix} = [T]^{-1} \begin{Bmatrix} \sigma_1 \\ \sigma_2 \\ \sigma_3 \\ \sigma_4 \\ \sigma_5 \\ \sigma_6 \end{Bmatrix}
\tag{3.15}
$$

The strains are transformable in the same manner:

$$
\begin{Bmatrix} \varepsilon_1 \\ \varepsilon_2 \\ \varepsilon_3 \\ \gamma_{12}/2 \\ \gamma_{13}/2 \\ \gamma_{23}/2 \end{Bmatrix} = [T] \begin{Bmatrix} \varepsilon_x \\ \varepsilon_y \\ \varepsilon_z \\ \gamma_{xy}/2 \\ \gamma_{xz}/2 \\ \gamma_{yz}/2 \end{Bmatrix}
\tag{3.16}
$$

Substituting Equation 3.16 into Equation 3.7 and then the result into Equation 3.15 we have:

$$
\begin{Bmatrix} \sigma_x \\ \sigma_y \\ \sigma_z \\ \sigma_{xy} \\ \sigma_{xz} \\ \sigma_{yz} \end{Bmatrix} = [T]^{-1}[C][T] \begin{Bmatrix} \varepsilon_x \\ \varepsilon_y \\ \varepsilon_z \\ \gamma_{xy}/2 \\ \gamma_{xz}/2 \\ \gamma_{yz}/2 \end{Bmatrix} = [C']\{\varepsilon'\}
\tag{3.17}
$$

The same is true for a 2 dimensional state of stress. Referencing Equation 3.11:

$$\left\{ \begin{array}{c} \sigma_x \\ \sigma_y \\ \sigma_{xy} \end{array} \right\} = [T]^{-1}[Q][T] \left\{ \begin{array}{c} \varepsilon_x \\ \varepsilon_y \\ \gamma_{xy}/2 \end{array} \right\} = [Q']\{\varepsilon'\} \qquad (3.18)$$

3.2.2 Elasto-Plastic Regime

Beyond the elastic range, there is no longer a one-to-one relationship between stress and strain and the material becomes load path dependent. The classical incremental theory of plasticity accounts for this loading history dependent behaviour while considering the effects of stress interaction under multiaxial stress states.

In essence, the incremental theory of plasticity relates the infinitesimal increment of plastic strain components ($d\epsilon^P_{ij}$) to the stress state (σ_{ij}) and the stress increment ($d\sigma_{ij}$). The constitutive relationships for the elasto-plastic regime are formulated through consideration of three fundamental concepts (Chen and Han, 1988): 1) the existence of an *initial yield surface*, 2) the *hardening rule* and 3) the *flow rule*. These concepts are explained in the following three sections.

3.2.2.1 Initial Yield Surface

Similar to the concept of a failure surface, the initial yield surface defines the combination of stress components (σ_i) at which elastic behaviour ends and plastic deformation begins. The quadratic function for yielding of an anisotropic plastic material is written in general form as (Shih and Lee, 1978)

$$f(\sigma_i, \alpha_i, M_{ij}, k) = 0 \tag{3.19}$$

In this study, the specific form of the yield function is taken to be

$$f = \overline{\sigma}^2(\sigma_i, \alpha_i, M_{ij}) - k^2 = 0 \tag{3.20}$$

where: $\overline{\sigma}$ is an *effective stress* or *equivalent stress* and k is a threshold stress which is equal to the size of the yield surface. For the purposes of plastic formulation, the Tsai-Wu criterion is adapted to this general form as follows:

The square of the effective stress is conveniently defined as

$$\overline{\sigma}^2 = M_{ij}(\sigma_i - \alpha_i)(\sigma_j - \alpha_j) \tag{3.21}$$

where the term M_{ij} describes the shape of the yield surface and $\alpha_i = \{\alpha_1, \alpha_2, \alpha_4\}^T = \{\alpha_1, \alpha_2, 0\}^T$ "describes the strength differential between the tensile and compressive strength or the 'offset' of the origin of the yield surface" (Shih and Lee, 1978), so that:

$$f \equiv M_{ij}(\sigma_i - \alpha_i)(\sigma_j - \alpha_j) - k^2 = 0 \qquad (i, j = 1, 2, 4) \tag{3.22a}$$

ie.,

$$f \equiv M_{11}\left[\sigma_1^2 - 2\sigma_1\alpha_1 + \alpha_1^2\right] + M_{22}\left[\sigma_2^2 - 2\sigma_2\alpha_2 + \alpha_2^2\right] +$$
$$2M_{12}\left[\sigma_1\sigma_2 - \sigma_1\alpha_2 - \sigma_2\alpha_1 + \alpha_1\alpha_2\right] + M_{44} \cdot \sigma_4^2 - k^2 = 0 \tag{3.22b}$$

Setting the equalities

$$L_i = 2 M_{ij}\alpha_j \quad ; \quad K = -M_{ij}\alpha_i\alpha_j + k^2 \qquad (3.23a \; ; \; 3.23b)$$

one obtains:

$$f \equiv M_{ij}\sigma_i\sigma_j - L_i\sigma_i - K = 0 \qquad (3.24)$$

Comparing Equation 3.24 with Equation 2.3, it is concluded that:

$$M_{ij} \equiv K F_{ij} = K \begin{bmatrix} F_{11} & F_{12} & 0 \\ F_{12} & F_{22} & 0 \\ 0 & 0 & F_{66} \end{bmatrix} \quad ; \quad L_i \equiv -K F_i = -K \begin{Bmatrix} F_1 \\ F_2 \\ 0 \end{Bmatrix} \qquad (3.25a; \; 3.25b)$$

It is noted that the components of M_{ij} are not independent. For example, if $M_{11} = 1$ then $K = 1/F_{11}$ and $M_{12} = F_{12}/F_{11}$, *etc...* (Shih and Lee 1978). Furthermore, α_i can be found by substituting Equation 3.25a into 3.23a, then equating the result to 3.25b

$$L_i = 2M_{ij}\alpha_j = 2KF_{ij}\alpha_j = -KF_i \qquad (3.26)$$

and solving the resulting simultaneous equations:

$$F_i = -2F_{ij}\alpha_j \qquad (3.27)$$

Finally, the square of the threshold stress is calculated from Equation 3.23b as

$$k^2 = K + M_{ij}\alpha_i\alpha_j \qquad (3.28)$$

When referring to initial yielding, Equation 3.20 is written in terms of the initial parameters, $\alpha_i{}^o$, $M_{ij}{}^o$, and k_o. ie.:

$$f \equiv \overline{\sigma}^2(\sigma_i, \alpha_i{}^o, M_{ij}{}^o) - k_o{}^2 = 0 \qquad (3.29)$$

3.2.2.2 Hardening Rule

Beyond the initial yield surface, material response is modeled according to inherent material behaviour. A material whose plastic deformation is assumed to occur under a constant yield stress level is termed perfectly plastic. For such a material, the yield surface remains that of Equation 3.29 throughout plastic flow. However, a material that exhibits stress dependency on plastic straining is said to undergo work or strain hardening and is modeled by specifying a new yield surface at each stage of plastic deformation known as subsequent yield surfaces. This post yield response, whereby the yield surface changes shape and/or position, is managed by way of a hardening parameter (χ) such that each new yield surface satisfies:

$$f \equiv \overline{\sigma}^2(\sigma_i, \alpha_i(\chi), M_{ij}(\chi)) - k^2(\chi) = 0 \qquad (3.30)$$

The hardening parameter relates the affected variables (α_i, M_{ij}, and k) to the degree of plastic deformation. In this study, 'χ' will be taken to be equivalent to a strain variable called the *effective plastic strain* $\bar{\varepsilon}_p$ (actually, $\int d\bar{\varepsilon}_p$). This variable is a scalar value function of the plastic strains.

Referencing Equation 3.30, one can establish the state of loading for a material under a multiaxial stress state. The criterion for a "loading state" (ie. when additional plastic deformation will occur) is

$$\text{if} \quad f = 0 \quad \text{and} \quad \frac{\partial f}{\partial \sigma_i} d\sigma_i > 0, \quad \text{then} \quad d\chi \neq 0 \tag{3.31}$$

and the stress point remains on an expanding yield surface. During "neutral loading"

$$\text{if} \quad f = 0 \quad \text{and} \quad \frac{\partial f}{\partial \sigma_i} d\sigma_i = 0, \quad \text{then} \quad d\chi = 0 \tag{3.32}$$

which is plastic loading for a perfectly plastic material and the stress point remains on the constant yield surface and finally, for a state of "unloading" we have

$$\text{if} \quad f = 0 \quad \text{and} \quad \frac{\partial f}{\partial \sigma_i} d\sigma_i < 0, \quad \text{then} \quad d\chi = 0 \tag{3.33}$$

and the stress point returns inside the yield surface.

Equation (3.30) represents a general anisotropic hardening model which specifies the rule for the evolution of the subsequent yield surfaces. Different hardening rules are formed by imposing different conditions on parameter variation with χ. If parameters α_i and M_{ij} are assumed to remain constant through plastic flow and k varies with χ, then the subsequent yield surfaces expand uniformly around the original yield surface, as in Figure 3.3, and the hardening model is said to be *isotropic*. If, on the other hand, both k and M_{ij} remain constant, while α_i vary with χ, the subsequent yield surfaces do not change shape but instead translate in stress space as a rigid body. This model is referred to as *kinematic hardening*. In the present study, a special type of general anisotropic hardening model is used whereby k varies proportionally with χ, α_i remain constant while the nonlinear yield parameters of M_{ij} vary non-proportionally with plastic deformation. This model leads to an expansion and at the same time distortion of the yield surface upon plastic strain. A similar approach was taken by Vaziri et al. (1991). This study, however, did not account for the difference in strength between tensile and compressive modes (ie. α_i was ignored). Under the present formulation, Equation 3.30 becomes

$$f \equiv \bar{\sigma}^2(\sigma_i, \alpha_i, M_{ij}(\chi)) - k^2(\chi) = 0$$
$$= M_{ij}(\chi)(\sigma_i - \alpha_i)(\sigma_j - \alpha_j) - k^2(\chi) = 0 \tag{3.34}$$

The mathematical details of this model are presented in Appendix A.

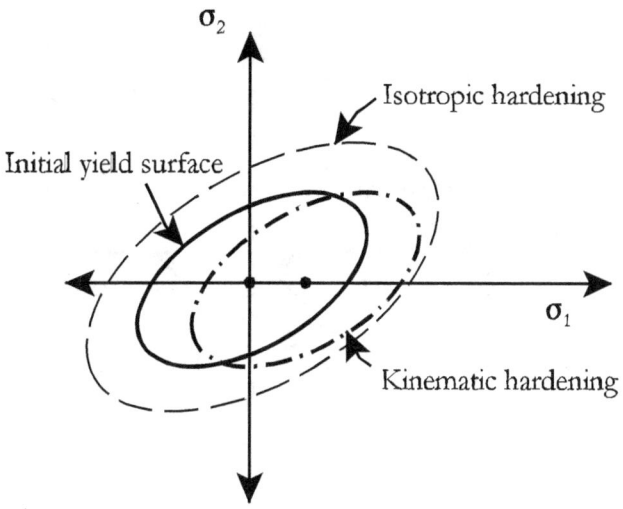

Figure 3.3 - Subsequent Yield Surfaces for Hardening Material

3.2.2.3 Flow Rule

In accordance with plasticity theory, under yielding, the total incremental strain vector is assumed to be composed of both incremental elastic and plastic components through simple superposition such that

$$d\varepsilon_i = d\varepsilon_i{}^e + d\varepsilon_i{}^p \tag{3.35}$$

The relationship between elastic strain increment and stress increment is defined through an infinitesimal form of Hooke's Law as

$$d\sigma_i = C_{ij}{}^e d\varepsilon_j{}^e \tag{3.36}$$

In what follows, the relationship between the total strain and stress increment beyond yield is derived. The flow rule enables a relationship to be made between the plastic strain component and the stress increment by introducing the concept of a *plastic potential function*, g (σ_i, α_i, M_{ij}, k). In particular, the *flow rule* is defined by classic plasticity theory as

$$d\varepsilon_i{}^p = d\lambda \frac{\partial g}{\partial \sigma_i} \tag{3.37}$$

where $d\varepsilon_i{}^p$ is the plastic strain increment vector, $d\lambda$ (termed the *plastic multiplier*) is the magnitude of the vector and $\partial g/\partial \sigma_i$, the stress gradient of the plastic potential function, defines the direction of the vector. For simplicity as well as computational efficiency, an associated flow rule will be assumed whereby the plastic potential function, g (σ_i, α_i, M_{ij}, k), is considered equal to the yield function, f (σ_i, α_i, M_{ij}, k) so that the direction of the plastic flow vector is normal to the yield surface as illustrated in Figure 3.4.

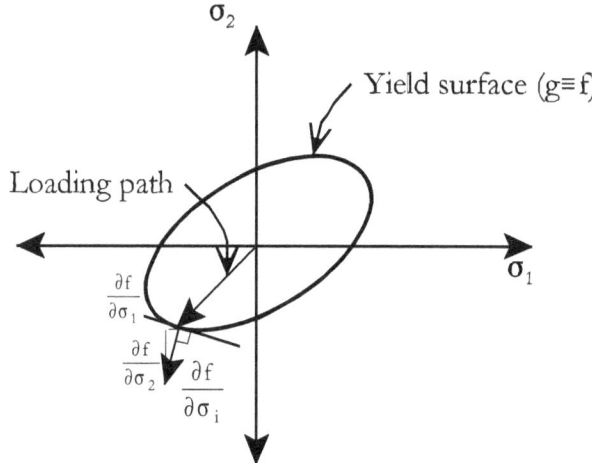

Figure 3.4 - Illustration of Normality of Plastic Flow Vector for Associated Plasticity

The hardening process of a material under a multiaxial stress state is characterized through one calibrated uniaxial stress-strain curve, using the effective stress ($\overline{\sigma}$) and the effective plastic strain ($\overline{\epsilon}^P$), which are scalar functions of the stresses and plastic strains respectively. The effective stress is defined by Equation 3.21. To define the effective plastic strain we consider a simple combination of infinitesimal plastic strain increments, the simplest type with the correct 'dimension' is given by Chan and Han (1988) as

$$(d\overline{\epsilon}^P)^2 = C \cdot d\epsilon_i^P \cdot d\epsilon_j^P \tag{3.38}$$

where C is a constant. If we choose $C = M_{ij}^{-1}$ (the inverse of M_{ij}, providing $|M_{ij}| \neq 0$), then we can simplify 3.38 using the flow rule to

$$
\begin{aligned}
(d\overline{\epsilon}^P)^2 &= M_{ij}^{-1} \cdot d\lambda^2 \cdot \frac{\partial f}{\partial \sigma_i} \frac{\partial f}{\partial \sigma_j} \\
&= M_{ij}^{-1} \cdot d\lambda^2 \cdot 2M_{ij}(\sigma_j - \alpha_j) \cdot 2M_{ik}(\sigma_k - \alpha_k) \\
&= d\lambda^2 \cdot 4M_{ij}(\sigma_i - \alpha_i) \cdot (\sigma_j - \alpha_j) \\
&= d\lambda^2 \cdot 4\sqrt{k}
\end{aligned}
\tag{3.39}
$$

Thus,

$$d\lambda = \frac{d\overline{\epsilon}^P}{2k} \tag{3.40}$$

From expression 3.34, we find that

$$\frac{\partial f}{\partial \sigma_i} = 2\overline{\sigma} \cdot \frac{\partial \overline{\sigma}}{\partial \sigma_i} = 2k \cdot \frac{\partial \overline{\sigma}}{\partial \sigma_i} \tag{3.41}$$

Designating $\dfrac{\partial \overline{\sigma}}{\partial \sigma_i} = a_i$, and substituting Equation 3.40 and 3.41 into the flow rule we have

$$
\begin{aligned}
d\varepsilon_i{}^P &= d\lambda \cdot 2k \cdot a_i \\
&= \dfrac{d\overline{\varepsilon}^P}{2k} \cdot 2k \cdot a_i \\
&= d\overline{\varepsilon}^P \cdot a_i
\end{aligned}
\tag{3.42}
$$

A one to one relationship between effective stress and effective plastic strain is assumed as follows:

$$
\overline{\sigma} = H(\overline{\varepsilon}^P)
\tag{3.43}
$$

where, upon differentiating

$$
\dfrac{d\overline{\sigma}}{d\overline{\varepsilon}^P} = H' \quad \text{or}, \quad \dfrac{dk}{d\overline{\varepsilon}^P} = H'
\tag{3.44a), (3.44b}
$$

H' is termed the hardening (or plastic) modulus and is associated with the rate of expansion of the yield surface. In the present study, a trilinear stress-strain curve is assumed such that H' is a constant. Thus, the yield surface expands in accordance with

$$
k = k_o + H'\overline{\varepsilon}^P
\tag{3.45}
$$

Substituting expression 3.35 into 3.44 while considering uniaxial yielding (typically in one of the principal material directions), we can determine the function H' as:

$$
H' = \dfrac{d\sigma}{d\varepsilon^P} = \dfrac{d\sigma}{d\varepsilon - d\varepsilon^P} = \dfrac{1}{\dfrac{d\varepsilon}{d\sigma} - \dfrac{d\varepsilon^P}{d\sigma}}
\tag{3.46}
$$

where, in the present study, we obtain the elastic stiffness, $E_{1c} = d\varepsilon/d\sigma$ and the tangent stiffness $E_{1c}' = d\varepsilon^P/d\sigma$ from the stress-strain curve of the compression parallel-to-grain test data. Thus, we have

$$
H' = \dfrac{1}{\dfrac{1}{E_{1c}} - \dfrac{1}{E_{1c}'}} = \dfrac{E_{1c}'}{1 - \dfrac{E_{1c}'}{E_{1c}}}
\tag{3.47}
$$

During plastic flow, each new state must satisfy the subsequent yield function so that

$$
f(\sigma_i, \alpha_i, M_{ij}, k) + df = 0
\tag{3.48}
$$

or, in this case

$$df = \frac{\partial f}{\partial \sigma_i} d\sigma_i + \frac{\partial f}{\partial M_{ij}} dM_{ij} + \frac{\partial f}{\partial k} dk = 0 \qquad (3.49)$$

which is known as the *consistency condition*. This condition assures that the subsequent stress and strain states remain on the subsequent yield surface. We consider each term, in turn.

Using Equations 3.41 and substituting 3.35 into 3.36, then considering 3.42, we have

$$\frac{\partial f}{\partial \sigma_i} d\sigma_i = 2k a_i \cdot d\sigma_i$$

$$= 2k a_i \cdot C_{ij}(d\varepsilon_j - d\varepsilon_j^P) \qquad (3.50)$$

$$= 2k a_i \cdot C_{ij}(d\varepsilon_j - a_j d\overline{\varepsilon}^P)$$

Next, referencing expression 3.34, noting $d\chi = d\overline{\varepsilon}^P$ and using Equation 3.44b

$$\frac{\partial f}{\partial M_{ij}} dM_{ij} = (\sigma_i - \alpha_i)(\sigma_j - \alpha_j) dM_{ij}$$

$$= (\sigma_i - \alpha_i)(\sigma_j - \alpha_j)\frac{\partial M_{ij}}{\partial \chi} d\chi$$

$$= (\sigma_i - \alpha_i)(\sigma_j - \alpha_j)\frac{\partial M_{ij}}{\partial \overline{\varepsilon}^P} d\overline{\varepsilon}^P \qquad (3.51)$$

$$= (\sigma_i - \alpha_i)(\sigma_j - \alpha_j)\frac{\partial M_{ij}}{\partial \overline{\varepsilon}^P}\frac{\partial \overline{\varepsilon}^P}{\partial k} dk$$

$$= (\sigma_i - \alpha_i)(\sigma_j - \alpha_j)\frac{\partial M_{ij}}{\partial k} H' d\overline{\varepsilon}^P$$

Referencing Equation 3.34 and 3.44b

$$\frac{\partial f}{\partial k} dk = -2k H' d\overline{\varepsilon}^P \qquad (3.52)$$

The consistency Equation 3.49 becomes:

$$df = \frac{\partial f}{\partial \sigma_i} d\sigma_i + \frac{\partial f}{\partial M_{ij}} dM_{ij} + \frac{\partial f}{\partial k} dk = 0$$

$$= 2k a_i C_{ij}^e (d\varepsilon_j - a_j d\overline{\varepsilon}^P) + \Pi \cdot H' d\overline{\varepsilon}^P - 2k H' d\overline{\varepsilon}^P = 0 \qquad (3.53)$$

where $\Pi = (\sigma_i - \alpha_i)(\sigma_j - \alpha_j)\frac{\partial M_{ij}}{\partial k}$.

Solving for $d\bar{\varepsilon}^p$, we have

$$d\bar{\varepsilon}^p = \frac{a_i C_{ij}^e}{H'\left(1 - \dfrac{\Pi}{2k}\right) + a_m C_{mn}^e a_n} d\varepsilon_j \qquad (3.54)$$

Finally, substituting this into Equation 3.42 and the result into Hooke's law $d\sigma_i = C_{ij}^e (d\epsilon_j - d\epsilon_j^p)$ the relationship between the incremental stress and total incremental strain beyond yield is

$$d\sigma_i = \left[C_{ij}^e - \frac{C_{ik}^e a_k a_l C_{lj}^e}{H'\left(1 - \dfrac{\Pi}{2k}\right) + a_m C_{mn}^e a_n} \right] d\varepsilon_j = \left[C_{ij}^e - C_{ij}^p \right] d\varepsilon_j = C_{ij}^{ep} d\varepsilon_j \qquad (3.55)$$

The plastic stiffness tensor, C_{ij}^p represents the degradation of the elastic stiffness tensor, C_{ij}^e due to plastic flow. A similar formulation for fiber-reinforced composite laminates is provided in Vaziri et al. (1991).

3.2.3 Post Failure Regime

Onset of material failure is determined through implementation of a failure surface which serves as an upper bound to all other loading surfaces and remains unchanged throughout loading. It is similar in form to Equation 3.34, ie.

$$f \equiv \bar{\sigma}^2(\sigma_i, \alpha_i, M_{ij}^u) - k_u^2 = 0 \qquad (3.56)$$

$$= M_{ij}^u (\sigma_i - \alpha_i)(\sigma_j - \alpha_j) - k_u^2 = 0$$

where the suffix 'u' denotes ultimate value. The values M_{ij}^u and k_u are material constants which come from ultimate strengths obtained from experimental tests in the principal material directions described in Chapter 5.

3.3 COMMENTARY ON PLANE STRESS FORMULATION

It is noted here that the same two dimensional form of the Tsai-Wu criterion is used for both the two dimensional and three dimensional formulation. The rationale behind using this form of the yield/failure criterion for the 3 dimensional model is the following:

a) The cases investigated in the current study (which reflect most applications for structural composite lumber) are assumed as being in a state of plane stress. Thus, a plane stress criterion was deemed adequate. In any case, the three dimensional model will produce estimates of the out-of-plane stresses, at which point, a judgement could be made as to whether or not the assumption of plane stress is correct.

b)   A thorough investigation of parameters required for the three dimensional form of the Tsai-Wu criterion is too time and cost prohibitive within the scope and realm of the current thesis. As the present study is the first investigation of the chosen modeling technique, the two-dimensional model was deemed to suffice. It was rationalized that, if found necessary from this thesis, a more extensive criterion could be used for future analysis.

4.1 INTRODUCTION

The finite element method is a well established numerical technique used for analyses of structures and continua (Cook et al. 1989; Zienkiewicz, 1971). Formulated in a computer program, the method is a powerful and versatile engineering tool with a wide range of applicability. For the present study, the finite element method provides a procedure for determining displacement and stress distribution throughout structural members considering nonlinear material behaviour.

The constitutive model outlined in the previous sections has been implemented into a stochastic, materially nonlinear finite element based FORTRAN 77 program with extended capacity to perform Monte Carlo simulations entitled COMAP (COMposite Analysis Program). The program was adapted from a 2 dimensional nonlinear program provided by Owen and Hinton (1980). Both a 2 dimensional and 3 dimensional version of the program have been developed and investigated.

This chapter systematically presents the general foundations of, as well as the underlying assumptions made in developing both versions of the program, the numerical implementation of the constitutive model of the previous chapter into the finite element analysis and finally, verification of both models through comparison with proven numerical or analytical solutions to given problems.

4.2 TWO DIMENSIONAL LINEAR FINITE ELEMENT MODEL

Very often in the analysis of laminates the lamina is assumed to be in a simple 2 dimensional state of stress (or plane stress). As such, it was decided that a 2 dimensional approach to strength prediction would first be investigated. It was reasoned that if successful, the 2 dimensional approach would provide a cheaper and faster solution than that of a 3 dimensional model. In this section we summarize the general finite element technique for analysing composites using 2 d formulation. This formulation is described in more detail in Ochoa and Reddy (1992).

4.2.1 Displacement-Based Element Representation

A structural member may be discretized and mathematically represented as a series of finite elements. The isoparametric formulation makes it possible to represent elements which are distorted as shown in Figure 4.1b. For the 2 dimensional model used in this thesis, plane bilinear isoparametric elements with 4 nodes and 2 degrees of freedom at each node are used.

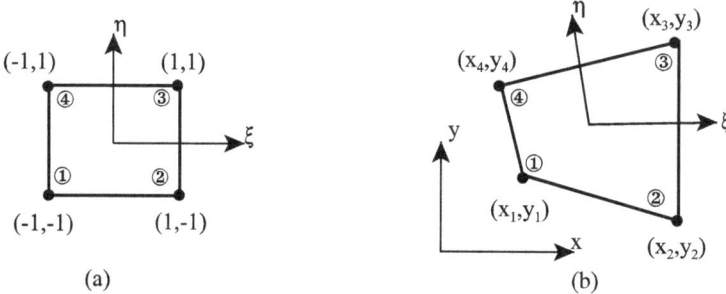

Figure 4.1 -   Plane Linear Isoparametric Element: (a) Rectangular element ; (b) Distorted element

While the four corners of the distorted element are defined by global Cartesian coordinates (x, y), displacements and coordinates are expressed in terms of curvilinear coordinates (ξ, η) through mapping. In keeping with isoparametric formulation, the displacements or the coordinates within the distorted element can be interpolated from nodal values as,

$$u = \sum_{i}^{n} N_i(\xi,\eta) \cdot u_i \quad ; \quad v = \sum_{i}^{n} N_i(\xi,\eta) \cdot v_i \qquad (4.1)$$

and

$$x = \sum_{i}^{n} N_i(\xi,\eta) \cdot x_i \quad ; \quad y = \sum_{i}^{n} N_i(\xi,\eta) \cdot y_i \qquad (4.2)$$

where, $N_i(\xi, \eta)$, known as *shape, basis or interpolation functions,* are chosen so as to preserve continuity of the displacements between elements. For this study, they are

$$N_i(\xi,\eta) = \frac{1}{4}\left(1 + \xi\xi_i\right)(1 + \eta\eta_i) \qquad (4.3)$$

where i is the number of the element nodal point, and ξ_i, η_i are the nodal coordinates in the ξ - η plane.

4.2.2 Stiffness Matrix Formulation

4.2.2.1 Compatibility

With a view to solving for the nodal displacements, the stiffness matrix for each element must first be obtained. Letting the elemental nodal displacements be written in vector form as $\{u\}^e = \{u_i, v_i\}^T$, the strains in the x , y coordinate system are defined using the strain-displacement relations as

$$
\left\{\begin{array}{c} \varepsilon_x \\ \varepsilon_y \\ \gamma_{xy} \end{array}\right\} = \left\{\begin{array}{c} \dfrac{\partial u}{\partial x} \\[6pt] \dfrac{\partial v}{\partial y} \\[6pt] \dfrac{\partial u}{\partial y}+\dfrac{\partial v}{\partial x} \end{array}\right\} = \begin{bmatrix} \dfrac{\partial N_1}{\partial x} & \dfrac{\partial N_2}{\partial x} & \dfrac{\partial N_3}{\partial x} & \dfrac{\partial N_4}{\partial x} & 0 & 0 & 0 & 0 \\[6pt] 0 & 0 & 0 & 0 & \dfrac{\partial N_1}{\partial y} & \dfrac{\partial N_2}{\partial y} & \dfrac{\partial N_3}{\partial y} & \dfrac{\partial N_4}{\partial y} \\[6pt] \dfrac{\partial N_1}{\partial y} & \dfrac{\partial N_2}{\partial y} & \dfrac{\partial N_3}{\partial y} & \dfrac{\partial N_4}{\partial y} & \dfrac{\partial N_1}{\partial x} & \dfrac{\partial N_2}{\partial x} & \dfrac{\partial N_3}{\partial x} & \dfrac{\partial N_4}{\partial x} \end{bmatrix} \left\{\begin{array}{c} u_1 \\ u_2 \\ u_3 \\ u_4 \\ v_1 \\ v_2 \\ v_3 \\ v_4 \end{array}\right\}
\tag{4.4}
$$

or,

$$
\left\{\varepsilon'\right\}^e = \begin{bmatrix} \dfrac{\partial N_i}{\partial x} & 0 \\[6pt] 0 & \dfrac{\partial N_i}{\partial y} \\[6pt] \dfrac{\partial N_i}{\partial y} & \dfrac{\partial N_i}{\partial x} \end{bmatrix} \left\{\begin{array}{c} u_i \\ v_i \end{array}\right\} = \left[B\right]\left\{u\right\}^e
\tag{4.4a}
$$

where [B] is designated as the strain displacement matrix. Here, the shape function derivatives are written with respect to the (x, y) coordinates, whereas the shape functions themselves are defined as functions of ξ and η. We can write by use of the general chain rule of differentiation

$$
\begin{aligned}
\frac{\partial N_i}{\partial \xi} &= \frac{\partial N_i}{\partial x}\frac{\partial x}{\partial \xi} + \frac{\partial N_i}{\partial y}\frac{\partial y}{\partial \xi} \\[6pt]
\frac{\partial N_i}{\partial \eta} &= \frac{\partial N_i}{\partial x}\frac{\partial x}{\partial \eta} + \frac{\partial N_i}{\partial y}\frac{\partial y}{\partial \eta}
\end{aligned}
\tag{4.5}
$$

such that the required derivatives $\partial N_i/\partial x$ and $\partial N_i/\partial y$ are obtained by inversion as

$$
\left\{\begin{array}{c} \dfrac{\partial N_i}{\partial x} \\[6pt] \dfrac{\partial N_i}{\partial y} \end{array}\right\} = \left[J^{-1}\right]\left\{\begin{array}{c} \dfrac{\partial N_i}{\partial \xi} \\[6pt] \dfrac{\partial N_i}{\partial \eta} \end{array}\right\}
\tag{4.6}
$$

where [J] is called the Jacobian matrix defined by

$$
[J] = \begin{bmatrix} \dfrac{\partial x}{\partial \xi} & \dfrac{\partial y}{\partial \xi} \\[6pt] \dfrac{\partial x}{\partial \eta} & \dfrac{\partial y}{\partial \eta} \end{bmatrix}
\tag{4.7}
$$

4.2.2.2 Equilibrium

Equilibrium of the structure is assured through application of the principle of virtual work whereby the internal virtual work (δW_{int}) is equal to the external virtual work (δW_{ext}), ie.:

$$\delta W_{int} = \int_V \{\delta\varepsilon'\}^T \cdot \{\sigma'\} dV \;=\; \delta W_{ext} = \int_V \{\delta u\}^T \cdot \{F\} dV \qquad (4.8)$$

where $\{\sigma'\} = \{\sigma_x, \sigma_y, \tau_{xy}\}^T$, $\{F\}$ are external body forces and $\{\delta u\}$ are the virtual displacements at the points of application of the forces. Considering Equation (4.4a) and simplifying, the equilibrium equation for an element is written in matrix form as

$$\int_{V^e} [B]^T \{\sigma'\}^e dV^e = \{F\}^e \qquad (4.9)$$

where $\{F\}^e$ is the equivalent external force acting on the nodal points.

4.2.2.3 Constitutive Relations

In order to resolve Equation 4.8, we must first consider the constitutive relationship between $\{\sigma'\}^e$ and $\{\varepsilon'\}^e$. In this 2 dimensional analysis, we employ classical lamination theory (CLT) to analyse the properties of the structural laminate.

Fundamental to the CLT, it is assumed that a state of plane stress exists in each layer; ie. the stresses in the through thickness (z) direction (reference Figure 2.1) are negligible. Also, the layers are assumed to be bonded perfectly together, so that there is no slip between plies. Furthermore, reflecting the conditions of this study, the analysis presented here is restricted to that of a symmetric laminate (with respect to its mid-plane), subjected to solely in-plane forces such that no bending or twisting moments exist. The consequence of using a symmetric laminate is that no coupling is assumed between bending and extension (ie. in-plane loads will not create bending or twisting curvature).

Referencing Figure 4.2, and using Equation 3.18, the stress-strain relationship for the kth lamina is expressed in terms of the mid-surface strain $\{\varepsilon'^o\}$ as

$$\{\sigma'\}^e_k = [Q']^e_k \{\varepsilon'\}^e_k = [Q']^e_k \{\varepsilon'^o\}^e \qquad (4.10)$$

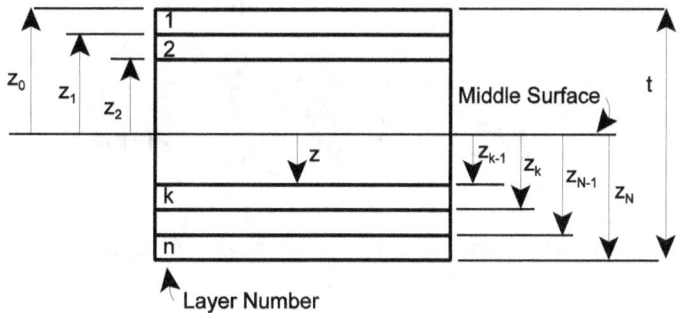

Figure 4.2 - Geometry of N - Layered Laminate (Jones, 1975)

The resultant laminate forces, $\{R\}$ measured per unit width, are obtained by integrating the stress components for each lamina over the entire laminate, ie.

$$\begin{Bmatrix} R_x \\ R_y \\ R_{xy} \end{Bmatrix} = \int_{-t/2}^{t/2} \begin{Bmatrix} \sigma_x \\ \sigma_y \\ \sigma_{xy} \end{Bmatrix}_k dz = \sum_{k=1}^{n} \left\{ \int_{z_{k-1}}^{z_k} \begin{Bmatrix} \sigma_x \\ \sigma_y \\ \sigma_{xy} \end{Bmatrix}_k dz \right\} \tag{4.11}$$

where t = laminate thickness, n = total number of layers, and z_k and z_{k-1} are as defined in Figure 4.2. Substituting Equation 4.9 into 4.10 and noting that the stresses are assumed constant through each lamina,

$$\{N\} = \sum_{k=1}^{n} \left[Q' \right]_k t_k \left\{ \varepsilon'^{\circ} \right\} \tag{4.12}$$

At this point, we can solve for the element stiffness matrix. Substituting Equation 4.10 into 4.9, we have

$$\int_{V^e} \left[B \right]^T \left[Q' \right]_k^e \left\{ \varepsilon'^{\circ} \right\}^e dV^e = \left\{ F \right\}^e \tag{4.13}$$

Further substitution of Equation 4.4a yields

$$\int_{V^e} \left[B \right]^T \left[Q' \right]_k^e \left[B \right] dV^e \cdot \left\{ u \right\}^e = \left\{ F \right\}^e$$

or,

$$\left[k \right]^e \cdot \left\{ u \right\}^e = \left\{ F \right\}^e \tag{4.14}$$

Thus, considering Equation 4.11, the element stiffness matrix is calculated as

$$\left[k \right]^e = \sum_{k=1}^{n} t_k \iint_{A^e} \left[B \right]^T \left[Q' \right]_k^e \left[B \right] dx\, dy \tag{4.15}$$

where the summation is taken over all layers in the laminate. Converting to curvilinear coordinates, we use the relation

$$dx\, dy = \det[J]\, d\xi\, d\eta \tag{4.16}$$

where $\det[J]$ is the determinant of the Jacobian, to get

$$[k]^e = \sum_{k=1}^{n} t_k \int_{-1}^{1}\int_{-1}^{1} [B]^T [Q']_k^e [B]\, \det[J]\, d\xi\, d\eta \tag{4.17}$$

The integrand in this equation is solved numerically using Gaussian quadrature with a 2 x 2 sampling scheme as follows:

$$[k]^e = \sum_{k=1}^{n}\sum_{j=1}^{2}\sum_{i=1}^{2} t_k [B]^T [Q']_k^e [B]\, \det[J]\, W_i W_j \tag{4.18}$$

where, in this case, the calculation is carried out at $\xi = \pm 1/\sqrt{3}$, $\eta = \pm 1/\sqrt{3}$ and W_i and W_j are associated weighting factors equal to 1.

The element stiffness matrices $[k]^e$ are added element-by-element, using standard assembly guidelines, to form the global stiffness matrix of the structure $[K]$. Knowing $[K]$, the applied loads and appropriate boundary conditions, the unknown nodal displacements for the entire structure are found from the simultaneous solution of the equations $[K] \cdot \{u\} = \{F\}$.

4.2.3 Two Dimensional Model Limitations

As noted in section 4.2.2.3, it is implicitly assumed in using 2 dimensional planar elements that interlaminar stresses associated with the through-thickness (z) direction are negligible. This is a valid assumption if 1) the laminate is symmetric, 2) the applied loads on the laminate are statically equivalent to in-plane forces producing neither bending nor twisting and 3) "free-edge" effects are negligible. The present study complies with the first two criteria. The third criteria leads to a phenomenon known as delamination which has been extensively studied both theoretically and experimentally for many advanced composite materials (Pipes and Pagano 1971; Whitney and Browning 1972; Gibson 1994). These studies show that within a "boundary region" (roughly one laminate thickness inward from the free edge of the laminate), a 3 dimensional stress state exists. Unfortunately, no such studies have been carried out on wood composites. It was decided that for further understanding of the material behaviour as well as providing a more versatile tool which could be used for more complex, non-symmetric composites, a 3 dimensional model would be investigated.

4.3 THREE DIMENSIONAL LINEAR FINITE ELEMENT MODEL

The 3 dimensional isoparametric procedure closely resembles the 2 dimensional development discussed in the previous section. For brevity, the following will concentrate primarily on the differences between the two models. This 3 dimensional finite element formulation is standard and can be found in any basic finite element text (Cook et al., 1989; Yang, 1986).

4.3.1 Displacement-Based Element Representation

The 3 dimensional element geometry is that of a linear solid (eight-node-brick) element as shown in Figure 4.3.

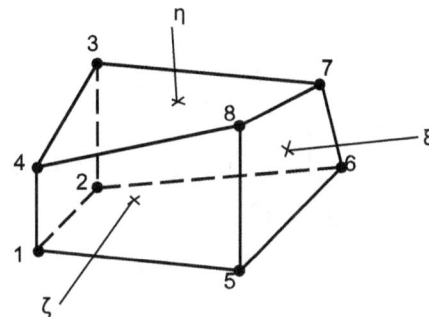

Figure 4.3 - Linear Solid Isoparametric Element

The displacements and the coordinates of a point are now defined in 3 directions as

$$u = \sum_i^n N_i(\xi,\eta,\zeta) \cdot u_i \quad ; \quad v = \sum_i^n N_i(\xi,\eta,\zeta) \cdot v_i \quad ; \quad w = \sum_i^n N_i(\xi,\eta,\zeta) \cdot w_i \tag{4.19}$$

and

$$x = \sum_i^n N_i(\xi,\eta,\zeta) \cdot x_i \quad ; \quad y = \sum_i^n N_i(\xi,\eta,\zeta) \cdot y_i \quad ; \quad z = \sum_i^n N_i(\xi,\eta,\zeta) \cdot z_i \tag{4.20}$$

The corresponding linear shape functions are found to be

$$N_i(\xi,\eta,\zeta) = \frac{1}{8}\left(1 + \xi\xi_i\right)(1 + \eta\eta_i)(1 + \zeta\zeta_i) \tag{4.21}$$

where ξ_i, η_i, and ζ_i are the coordinate values of node i, with i = 1, 2, ... 8.

4.3.2 Stiffness Matrix Formulation

4.3.2.1 Compatibility

The strain displacement relationship in 3 dimensions is written as

$$\{\varepsilon'\}^e = \left\{ \begin{array}{c} \varepsilon_x \\ \varepsilon_y \\ \varepsilon_z \\ \gamma_{xy} \\ \gamma_{xz} \\ \gamma_{yz} \end{array} \right\}^e = \begin{bmatrix} \dfrac{\partial N_i}{\partial x} & 0 & 0 \\[2mm] 0 & \dfrac{\partial N_i}{\partial y} & 0 \\[2mm] 0 & 0 & \dfrac{\partial N_i}{\partial z} \\[2mm] \dfrac{\partial N_i}{\partial y} & \dfrac{\partial N_i}{\partial x} & 0 \\[2mm] \dfrac{\partial N_i}{\partial z} & 0 & \dfrac{\partial N_i}{\partial x} \\[2mm] 0 & \dfrac{\partial N_i}{\partial z} & \dfrac{\partial N_i}{\partial y} \end{bmatrix} \left\{ \begin{array}{c} u_i \\ v_i \\ w_i \end{array} \right\} = [B]\{u\}^e \qquad (4.22)$$

Because N_i are defined in terms of ξ, η, and ζ, we use the inverse of the Jacobian matrix

$$[J] = \begin{bmatrix} \dfrac{\partial x}{\partial \xi} & \dfrac{\partial y}{\partial \xi} & \dfrac{\partial z}{\partial \xi} \\[2mm] \dfrac{\partial x}{\partial \eta} & \dfrac{\partial y}{\partial \eta} & \dfrac{\partial z}{\partial \eta} \\[2mm] \dfrac{\partial x}{\partial \zeta} & \dfrac{\partial y}{\partial \zeta} & \dfrac{\partial z}{\partial \zeta} \end{bmatrix} \qquad (4.23)$$

to find the derivatives of the shape functions in terms of x, y and z, as

$$\left\{ \begin{array}{c} \dfrac{\partial N_i}{\partial x} \\[2mm] \dfrac{\partial N_i}{\partial y} \\[2mm] \dfrac{\partial N_i}{\partial z} \end{array} \right\} = \left[J^{-1} \right] \left\{ \begin{array}{c} \dfrac{\partial N_i}{\partial \xi} \\[2mm] \dfrac{\partial N_i}{\partial \eta} \\[2mm] \dfrac{\partial N_i}{\partial \zeta} \end{array} \right\} \qquad (4.24)$$

4.3.2.2 Equilibrium

By applying the principle of virtual work, the equilibrium equation for an element is the same as for the 2 dimensional case,

$$\int_{V^e} [B]^T \{\sigma'\}^e dV^e = \{F\}^e \qquad (4.25)$$

4.3.2.3 Constitutive Relations

From Equation 3.17, the stress-strain relationship for each element in the (x,y,z) laminate coordinate system (see Figure 2.1) is defined as

$$\{\sigma'\}^e = [C']^e \{\varepsilon'\}^e \tag{4.26}$$

Substituting Equation 4.26 into 4.25, we have

$$\int_{V^e} [B]^T [C']^e \{\varepsilon'\}^e dV^e = \{F\}^e \tag{4.27}$$

By further substitution of the strain displacement relationship (Equation 4.22) into Equation 4.27,

$$\int_{V^e} [B]^T [C']^e [B] dV^e \cdot \{u\}^e = \{F\}^e \tag{4.28}$$

we find the element stiffness matrix in the usual manner as

$$[k]^e = \int_{V^e} [B]^T [C']^e [B] \, dx \, dy \, dz \tag{4.29}$$

Converting to curvilinear coordinates, we use the relation

$$dx \, dy \, dz = \det[J] d\xi \, d\eta \, d\zeta \tag{4.30}$$

to get

$$[k]^e = \int_{-1}^{1}\int_{-1}^{1}\int_{-1}^{1} [B]^T [C']^e [B] \det[J] d\xi \, d\eta \, d\zeta \tag{4.31}$$

Finally, using Gaussian quadrature with a 2 x 2 sampling scheme:

$$[k]^e = \sum_{k=1}^{2}\sum_{j=1}^{2}\sum_{i=1}^{2} [B]^T [C']^e [B] \det[J] W_i \, W_j \, W_k \tag{4.32}$$

The calculation is carried out at $\xi = \pm 1/\sqrt{3}$, $\eta = \pm 1/\sqrt{3}$, $\zeta = \pm 1/\sqrt{3}$ and W_i, W_j and W_k are the associated weighting factors equal to 1.

As in the 2 dimensional model, the element stiffness matrices $[k]^e$ are assembled to form the global stiffness matrix of the structure [K]. This matrix relates the nodal displacements for the entire structure with the applied loads through the relationship,

$$\{F\} = [K]\{u\} \tag{4.33}$$

4.4   NONLINEAR FORMULATION

In a nonlinear analysis, the governing equation is 4.25. This equation represents the equilibrium of the externally applied force vector $\{F\}$ with an internal resisting force vector $\{P\} = \int [B]^T \{\sigma'(\bar{u})\} dV$ which is a (materially) nonlinear function of the nodal displacements $\{u\}$. The difference between the two vectors can be interpreted as a load imbalance. Typically, the finite element solution proceeds on an incremental basis and solution of these equations requires an incremental/iterative numerical algorithm which effectively minimizes, to an acceptable tolerance, the residual

$$\{\Psi(\bar{u})\} = \{F\} - \{P(\bar{u})\} \tag{4.34}$$

There are numerous methods for solution of the nonlinear system of equations 4.34. The method used in this study is a modification of the classical Newton-Raphson method.

4.4.1   Modified Newton-Raphson Method

As found in any standard text on nonlinear finite element formulation (Owen and Hinton, 1980; Cook et al. 1989), we first consider a truncated Taylor series expansion of $\{\Psi\}_r$ about $\{u\}_{r-1}$:

$$\{\Psi\}_r = \{\Psi\}_{r-1} + \frac{\partial \{\Psi\}}{\partial \{u\}}\bigg|_{\{u\}_{r-1}} \cdot \{\Delta u\} \tag{4.35}$$

where the subscript r denotes the iteration number and $\{\Delta u\} = \{u\}_r - \{u\}_{r-1}$. Referencing Equation 4.34, it is noted that

$$[K]_{r-1} = \frac{\partial \{P\}}{\partial \{u\}}\bigg|_{\{u\}_{r-1}} = -\frac{\partial \{\Psi\}}{\partial \{u\}}\bigg|_{\{u\}_{r-1}} \tag{4.36}$$

where $[K]_{r-1}$ is the global stiffness matrix evaluated at the beginning of the rth iteration. Substituting this into Equation 4.35 and considering that ultimately we want $\{\Psi\}_r \cong 0$, we have

$$\{\Psi\}_{r-1} = [K]_{r-1} \cdot \{\Delta u\} \tag{4.37}$$

We use this set of equations to compute $\{\Delta u\}$ whereupon we can update the displacement estimate to

$$\{u\}_r = \{u\}_{r-1} + \{\Delta u\} \tag{4.38}$$

The new found displacements are used to calculate a new internal resisting force vector $\{P\}$ whereby we re-evaluate the residual $\{\Psi\}$ (equation 4.34). For the next iteration, we obtain a new structural (tangent) stiffness matrix $[K]$ and repeat the process until the residual converges to the point of satisfying a prescribed tolerance. In the present study, we have adopted an initial stiffness routine (illustrated in Figure 4.4 for a single variable situation), in which the tangent stiffness corresponds always to the initial value to avoid zero or negative definite stiffness beyond the peak load. This also reduces the costs associated with reproducing the tangent stiffness matrix for each iteration. However, more iterative cycles are needed within each iteration to reach the set tolerance.

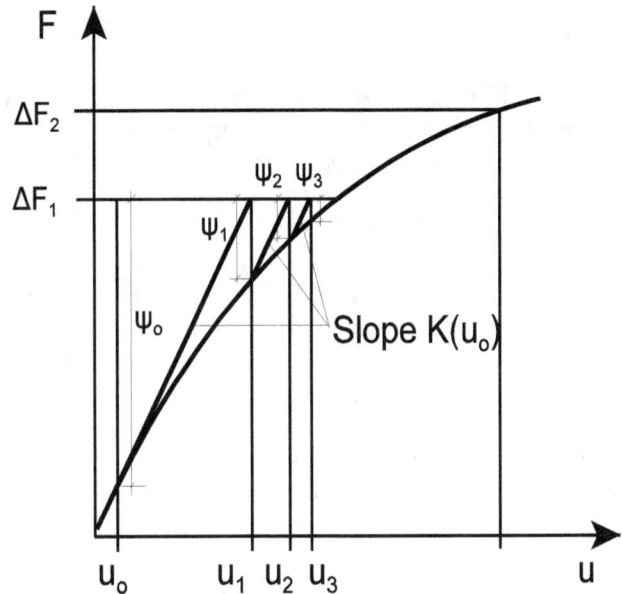

Figure 4.4 -   **Initial Stiffness Solution Algorithm for a Single Variable Situation (Owen and Hinton, 1980)**

The convergence criterion used is one based on the residual force values (Owen and Hinton, 1980):

$$\frac{\sqrt{\{\Psi\}_r^T \{\Psi\}}}{\sqrt{\{F\}^T \{F\}}} \leq \text{tolerance} \tag{4.39}$$

For this study, the tolerance was set to be 0.01.

4.4.2 Elastic-Plastic Formulation

As alluded in the previous section, the equations of equilibrium are resolved in an incremental manner. During any iteration within an increment in load (or displacement), a part of an element may yield or fail. To determine this, all stress and strain quantities are monitored at each Gaussian point throughout the entire stress history of the member. The purpose of this section is to detail the procedure used in monitoring and adjusting the stresses and strains throughout the elastic-plastic regime.

Yielding or failure of a point is established through use of the Tsai-Wu criterion. In the criterion, stresses are defined with respect to the principal material coordinate system. Thus, the stresses are transformed from the x-y coordinate system in accordance with the transformation relationships established in Chapter 3, before being used in the criterion. In the following discussion, stresses are denoted $\{\sigma\} = \{\sigma_1, \sigma_2, \sigma_4\}^T$ or $\{\sigma_1, \sigma_2, \sigma_3, \sigma_4, \sigma_5, \sigma_6\}^T$ for the 2-dimensional model or 3-dimensional model, respectively. Similarly, strains in the principal material directions are denoted, $\{\varepsilon\}$.

We follow the perspective of one Gaussian point within the member during the rth iteration. All stress and strain quantities associated with that point as well as the state of the yield surface are known from the $(r-1)^{th}$ iteration.

1. Using the residual forces from the previous iteration (ie. $\{\Psi\}_{r-1}$) which is the load imbalance, we calculate the resulting displacement (Equation 4.37) and from this, strain increment $\{\Delta\varepsilon\}$ (Equation 4.4a (2-D) or 4.22(3-D)).
2. Assuming elastic behaviour, we compute the elastic increment in stress $\{\Delta\sigma^e\}$ from Equation 3.7 (3-D) or 3.11 (2-D).
3. We calculate a trial stress, $\{\sigma^e\}_r$ which is the accumulated total stress for the point according to

$$\left\{\sigma^e\right\}_r = \{\sigma\}_{r-1} + \left\{\Delta\sigma^e\right\}_r \qquad (4.40)$$

4. We use our trial stress to calculate the effective stress $\bar{\sigma}_r = (M_{ij}^{\,o}(\sigma_i - \alpha_i)(\sigma_j - \alpha_j))^{1/2}$ and check this against the initial yield criterion (Equation 3.29); ie. we check if $\bar{\sigma}_r \geq k_o$. If the answer is no, then the elastic assumption is valid and the trial stress is indeed correct. If the answer is yes, however, the point has either a) yielded and the point continues to deform ductilely (as a result of being *compression - dominant*) or b) failed in a brittle manner (as a result of being *tension - dominant*).

Tension vs. Compression Dominance

The decision between compression or tension dominance is made depending on the combination of stresses at the point of failure. We define conditions for tension dominance for the two and three dimensional models as shown in Table 4.1. To describe these conditions, we need to categorize the individual components of the effective stress (referencing Equation 3.22b) as:

$$\begin{aligned}
\rho_1 &= M_{11}\left[\sigma_1^{\,2} - 2\sigma_1\alpha_1 + \alpha_1^{\,2}\right] \\
\rho_2 &= M_{22}\left[\sigma_2^{\,2} - 2\sigma_2\alpha_2 + \alpha_2^{\,2}\right] \\
\rho_4 &= M_{44} \cdot \sigma_4^{\,2}
\end{aligned} \qquad (4.41)$$

These components (ρ_1, ρ_2, and ρ_4) are chosen to reflect the magnitude of the stresses parallel-to-grain, perpendicular-to-grain and in-plane shear, respectively.

Table 4.1 - Conditions for Tension Dominance

2 D Model	3 D Model
$\sigma_1 \geq X_t$ $\sigma_2 \geq Y_t$ $\lvert \sigma_6 \rvert \geq S$ $\sigma_1 \geq 0$ and $\rho_1 \geq \rho_2$ $\sigma_2 \geq 0$ and $\rho_2 \geq \rho_1$	$\sigma_1 \geq X_t$ $\sigma_2 \geq Y_t$ $\lvert \sigma_6 \rvert \geq S$ $\sigma_1 \geq 0$ and $\rho_1 \geq \rho_2$ $\sigma_2 \geq 0$ and $\rho_2 \geq \rho_1$ $\rho_4 \geq \rho_1$ and $\rho_4 \geq \rho_2$ $\left\lvert \dfrac{\sigma_4}{S} \right\rvert \geq \left\lvert \dfrac{\sigma_1^+}{X_t} \right\rvert \; or \; \left\lvert \dfrac{\sigma_1^-}{X_c} \right\rvert$

At an integration point, the stress state is deemed tension-dominant, and thereby brittle, when the failure criterion is violated and any one of the conditions shown occurs; otherwise, it is deemed compression-dominant. The rationale as to whether the point yields or fails is albeit subjective. It was chosen to provide good results when compared to experimental data. Tensile failure leads to the post-failure regime and will therefore be deferred for discussion in the next section. Stress combinations other than that defined by Table 4.1 result in ductile behaviour and follow plasticity formulation. We continue with elastic-plastic formulation.

Given that the trial stress has been determined to be compression dominant and $\overline{\sigma}_r \geq k_0$, then the stress vector has crossed the initial yield surface. This stress path is illustrated vectorially in Figure 4.5 (Owen and Hinton, 1980).

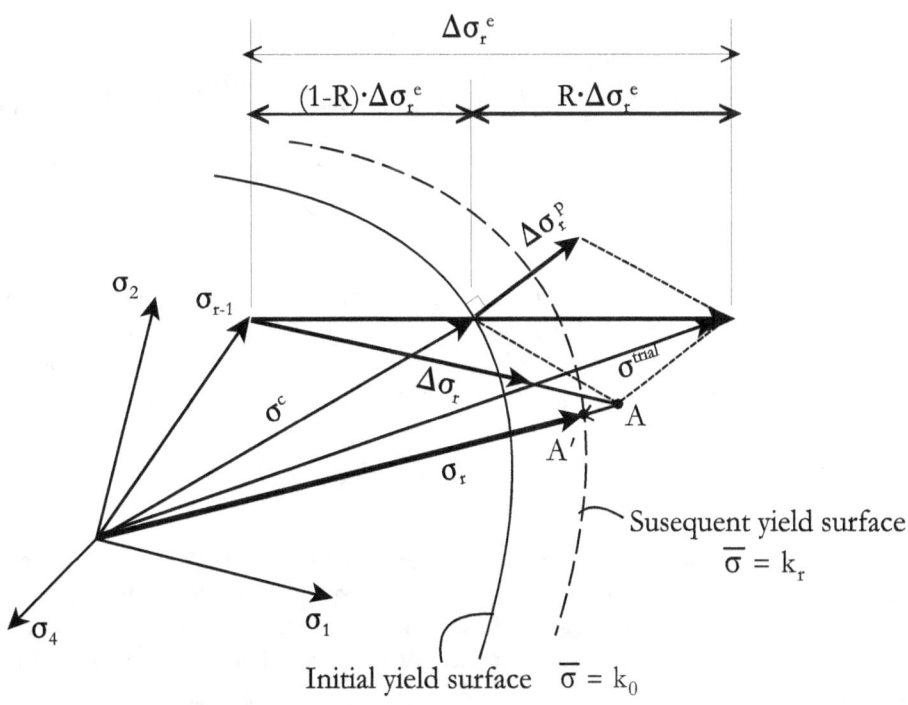

Figure 4.5 - Incremental Stress Changes at Initial Yield

5. For yielded Gauss points, we wish to proportion the total stress appropriately into elastic and elasto-plastic portions. For this, we first compute a contact stress $\{\sigma^c\}$ (entirely elastic) where the stress path just comes into contact with the initial yield surface, ie.,

$$\{\sigma^c\} = \{\sigma\}_{r-1} + (1 - R)\{\Delta\sigma^e\} \tag{4.42}$$

where, R is determined through simple linear interpolation:

$$R = \frac{\overline{\sigma}_r^e - k_o}{\overline{\sigma}_r^e - \overline{\sigma}_{r-1}} \tag{4.43}$$

The remaining portion of the elastic stress increment, $R \cdot \{\Delta\sigma^e\}$ lays beyond the initial yield surface (which violates the yield criterion) and must be corrected to account for plastic deformation. It is noted that if the Gaussian point had previously yielded, then $R = 1$ and the entire stress state is elastoplastic.

6. Referring to Figure 4.5, we wish to find the total stresses $\{\sigma_r\}$ which satisfy elasto-plastic conditions. Substituting incremental changes for infinitesimal changes in Equation 3.55 we have

$$
\begin{aligned}
\{\Delta\sigma_r\} &= \left[C^{ep}\right]\{\Delta\varepsilon_r\} \\
&= \left[C^e\right]\{\Delta\varepsilon_r\} - \left[C^p\right]\{\Delta\varepsilon_r\} \\
&= \{\Delta\sigma_r^e\} - \{\Delta\sigma_r^p\}
\end{aligned} \tag{4.44}
$$

Thus, when incrementing from $\{\sigma_{r-1}\}$, we can calculate

$$\{\sigma_r\} = \{\sigma_{r-1}\} + \{\Delta\sigma_r^e\} - \frac{\left[C^e\right]\{a\}\{a\}^T\left[C^e\right]}{H'\left(1 - \dfrac{\Pi}{2k}\right) + \{a\}^T\left[C^e\right]\{a\}}\{\Delta\varepsilon_r\} \tag{4.45}$$

where all values are as defined in Section 3.2.2. The final stress point, corresponding to point A on the schematic, may not fall directly on the yield surface (ie. at A′). The discrepancy is aggravated with larger increments however can be resolved by simply scaling the individual stress components of $\{\sigma_r\}$. As the effective stress, $\overline{\sigma}$ should coincide with the threshold stress, k, an appropriate scaling factor is found to be

$$\{\sigma_r\} = \{\sigma_r\} \cdot \left(\frac{k}{\overline{\sigma}_r}\right) \tag{4.46}$$

If large increments in load (or displacement) is allowed in the preceding process, the final point A′ may contain some error. As explained in Owen and Hinton (1980), the inaccuracy may be reduced by relaxing the excess stress to the yield surface in several stages. Referencing Figure 4.6, the excess stress is divided into m parts where m is the closest integer less than

$$\left(\frac{\overline{\sigma}_r^e - k}{k_o}\right) \cdot 8 + 1 \tag{4.47}$$

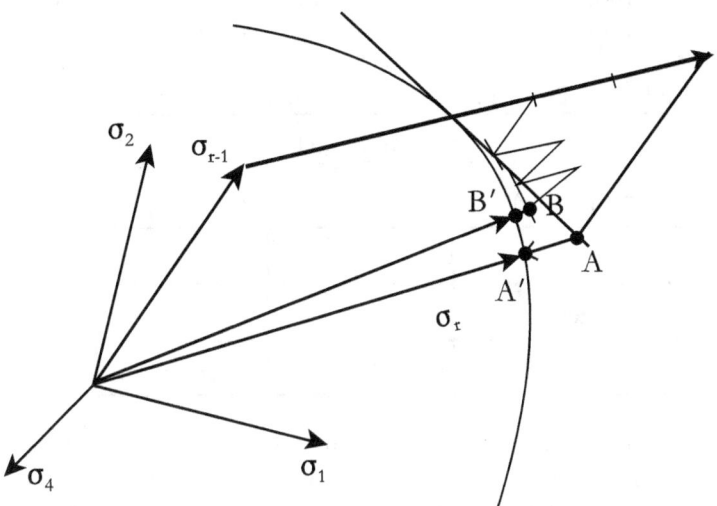

Figure 4.6 - Stepped Process to Reduce Stress Point to Yield Surface

Using a trilinear representation of the stress-strain curve, the threshold stress, k is updated after each iteration according to Equation 3.45: $k_r = k_0 + H' \varepsilon^P{}_r$. In accordance with the anisotropic hardening rule adopted in this study, the nonlinear strength variables, X_c and Y_c are also updated after each iteration, as described in Appendix A.

7. Finally, from the total stresses established for each Gaussian point, whether in an elastic state or an elastoplastic state, appropriately transformed back to the x - y coordinate system, we compute the equivalent nodal forces

$$\{P\}_r = \int_V [B]^T \{\sigma'_r\} dV \qquad (4.48)$$

which are used in calculating the residual forces $\{\Psi\}$ for the next iteration, if deemed necessary by the convergence criterion Equation 4.39.

This iterative process repeats for each load increment until the stresses reach the ultimate strengths per Equation 3.56. At this point, failure occurs.

4.4.3 Post Failure Regime

When failure has been established, post-failure behaviour ensues. Failure is described as being either brittle or ductile, depending on whether the combination of stresses at the point of failure result in a tension or compression dominant state (see section 4.4.2). Referencing Figure 3.1, when ductile failure occurs, the yield quantities, X_c and Y_c have reached their ultimate values and the stress level remains constant, reflecting that the material cannot support additional stress. Stresses are not permitted to move outside the ultimate yield surface and thus can only traverse the surface until both equilibrium and the constitutive relations are satisfied. When brittle failure occurs, however, the stress level is reduced gradually to facilitate convergence of the iterative procedure. The plane stresses $\{\sigma_1, \sigma_2, \sigma_4\}$ are reduced proportionate to the same values in the previous load increment. The ratios used differ slightly for the 2-dimensional and 3-dimensional model and are summarized in Table 4.2. These values were established through comparison with test results to provide good post-peak results.

Table 4.2 - Brittle Failure Stress Reduction Ratios

Stress	2 - d Model	3 - d Model
σ_1^+ (tensile)	0.70	0.90
σ_1^- (compressive)	0.98	0.98
σ_2^+ (tensile)	0.90	0.90
σ_4	0.95	0.95

In summary, when a local integration point is detected to have failed either in a brittle or ductile manner, the global material stiffness is modified until satisfaction of equilibrium is attained within the iterative process. Failure of a portion of one lamina is compensated for by an increase in the load carried by adjacent Gaussian points or laminae. The load is then further increased until final failure is reached. Final failure occurs when the load displacement curve begins to descend and the program is automatically halted when the load reaches 90 percent of the peak load. In rare cases, the program may also be halted due to nonconvergence beyond peak load. The 90 percent condition was chosen both in consideration of computer time as well as numerical instability.

4.5 PROGRAM VERIFICATION

The theory presented in the foregoing chapters has been implemented into a finite element based computer program, COMAP. As with any new program, however, a thorough check is initially necessary before it can be used with confidence. It is the objective of this section to verify the general performance and results of COMAP. As no other software exists which has the same capabilities as our program, the verification process involves comparison with several programs -each targeting a specific aspect of COMAP. As well, a problem with an exact analytical solution is investigated to gauge the robustness of the program's prediction capabilities (within the bounds of the theoretical assumptions).

4.5.1 Isotropic Cantilever Analysis

In order to assess the plastic deformation capabilities of the program, the classic example of an inelastic cantilever beam, shown in Figure 4.7, is investigated.

The example considers the case of a elastic - perfectly plastic material (such as structural steel), with rectangular cross section, limited to small deflections and neglecting the effects of shear on deflection. The problem has known analytical solutions for both elastic and elastic-plastic behaviour (Timoshenko, 1972). For the elastic range, it is known that the yield load F_y which first produces yielding of the beam is given by

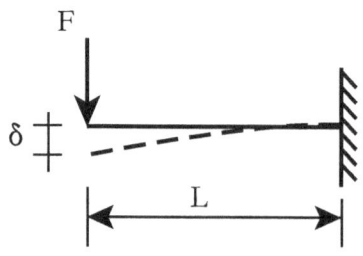

Figure 4.7 - Cantilever Beam Example

$$F_y = \frac{M_y}{L} \qquad\qquad (4.49)$$

The deflection at incipient yield is

$$\delta_y = \frac{F_y L^3}{3EI} \qquad\qquad (4.50)$$

In nondimensional form, we have a simple relationship between deflection and load in the entire elastic range,

$$\frac{\delta}{\delta_y} = \frac{F}{F_y} \qquad \left(0 \le \frac{F}{F_y} \le 1 \right) \qquad\qquad (4.51)$$

In the region of elastic-plastic behaviour, the relationship is shown to be

$$\frac{\delta}{\delta_y} = \left(\frac{F_y}{F}\right)^2 \left[5 - \left(3 + \frac{F}{F_y}\right)\sqrt{3 - \frac{2F}{F_y}} \right] \qquad \left(1 \le \frac{F}{F_y} \le \frac{3}{2} \right) \qquad\qquad (4.52)$$

A graph of load F/F_y versus deflection δ/δ_y through the entire stress path is shown in Figure 4.8.

This result forms the basis of comparison for the numerical analysis using the program COMAP. The load displacement curve is reproduced using both the 2-dimensional and 3-dimensional version of the program assuming an isotropic material. The cantilever beam is discretized for both cases as shown in Figure 4.9.

Linear 4-node, or 8-node elements are used as shown, together with a 2x2 (or 2x2x2) Gauss integration scheme. Load was applied incrementally at the free end of the beam lumped proportionally on the nodal points. The assumed isotropic material properties are summarized as follows:

Elastic Modulus:	$E = 30 \times 10^3$ MPa
Poisson's Ratio:	$\nu = 0.3$
Uniaxial Yield Stress:	$\sigma_y = 30$ MPa
Hardening Modulus:	$H' = 0$

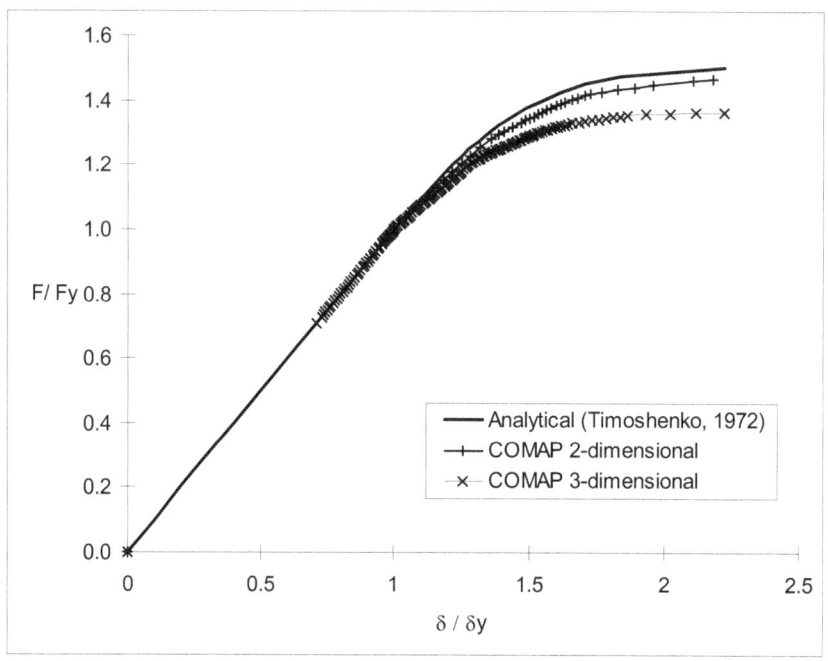

Figure 4.8 - Load versus Displacement for Isotropic Cantilever Beam Example

As the program is configured for a stochastic analysis with regards to the elastic moduli and yield stresses, the values aforementioned are considered to be the average values and standard deviations are zero. The load F was incremented until plastic collapse of the beam was attained. Collapse was deemed to have occurred when the iterative procedure diverged for an incremental load increase. The load increments are shown as crosses in Figure 4.8. Comparing the three curves, the program reproduces the analytical results reasonably well. The slight discrepancy in the elastic-plastic region, specifically for the 3-dimensional program, is attributed to the fact that the failure criterion does not take into consideration the stresses in the through-thickness direction. Also, the stresses were evaluated at the Gaussian integration points. Evaluating the conditions at the extreme fiber instead (where the longitudinal stresses would be approximately 12 percent higher) would lead to a lower yield load and thus a closer approximation of the analytical result.

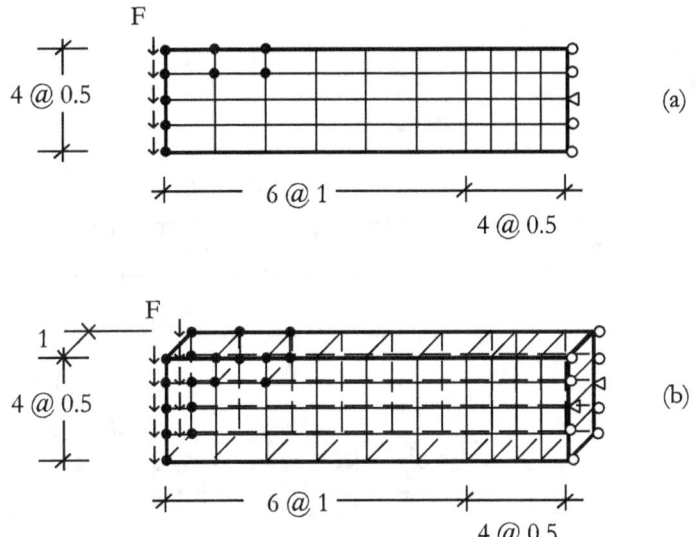

Figure 4.9 - Finite Element Mesh for a) 2 and b) 3 Dimensional Analysis of Isotropic Cantilever Beam

4.5.2 5 Point Continuous Beam Analysis

A further verification of the program entailed a comparison of the displacement and stress results obtained from a common commercial finite element program, ANSYS®, with the results obtained from COMAP. The purpose of this analysis is to confirm the accuracy of the linear elastic, *orthotropic* finite element formulation. The problem under consideration is that of a continuous beam illustrated in Figure 4.10.

As the beam is symmetric with respect to the longitudinal axis, only one half of the physical problem is modeled. For both programs, COMAP and ANSYS® in 2 and 3 dimensions, the geometrical parameters and finite element mesh are as shown in Figure 4.11. The applied load was chosen such that the resultant internal stresses remained within the elastic range.   A total load of approximately 360 N/mm width was applied, as shown. Linear 4-node, or 8-node elements were used, together with a 2x2 (or 2x2x2) Gauss integration scheme for both programs.

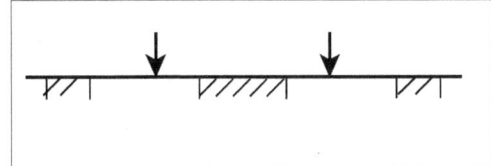

Figure 4.10 - 5 Point Continuous Beam

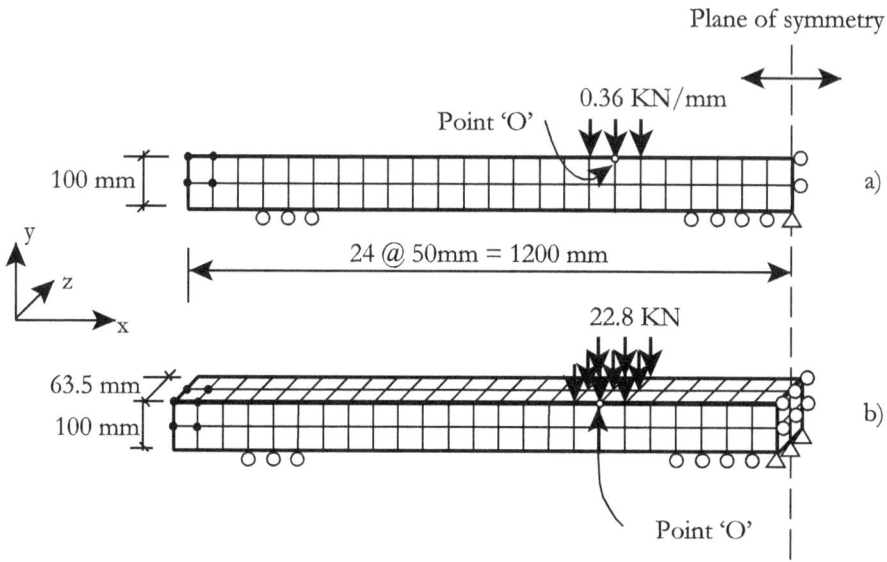

Figure 4.11 -   Finite Element Mesh for Continuous Beam Example: a) 2 dimensions b) 3 dimensions

The deterministic elastic properties for the material are:

Longitudinal Elastic Modulus: $E_x = 11 \times 10^3$ MPa
Transverse Elastic Modulus: $E_y = 0.4 \times 10^3$ MPa
Through-thickness Elastic Modulus: $E_z = 0.62 \times 10^3$ MPa (for 3 d. analysis only)
Poisson's Ratios: $\nu_{xy} = 0.32$; $\nu_{yx} = 0.012$; (for 2 d. analysis)
$\nu_{xz} = 0.29$; $\nu_{zx} = 0.01$; $\nu_{yz} = 0.20$ and $\nu_{zy} = 0.20$ (for 3 d. analysis only)
Shear Moduli: $G_{xy} = 0.7 \times 10^3$ MPa (for 2 d. analysis)
$G_{xz} = 0.76 \times 10^3$ MPa; $G_{yz} = 0.08 \times 10^3$ MPa (for 3 d. analysis only)

Poisson's ratios were calculated based on ν_{xy} and the Elastic moduli in accordance with Equation 3.10. The general behaviour of the beam according to each program was observed. The longitudinal stress (σ_x) contours, as well as deflection behaviour are demonstrated in Figure 4.12. For both the 2-dimensional and 3-dimensional cases, the COMAP results agreed favorably with that for ANSYS®. The results for the specific point 'O' (depicted in Figure 4.11) is provided in Table 4.3. Any discrepancies are minor and are attributed primarily to extrapolation of stresses from Gaussian points to nodes.

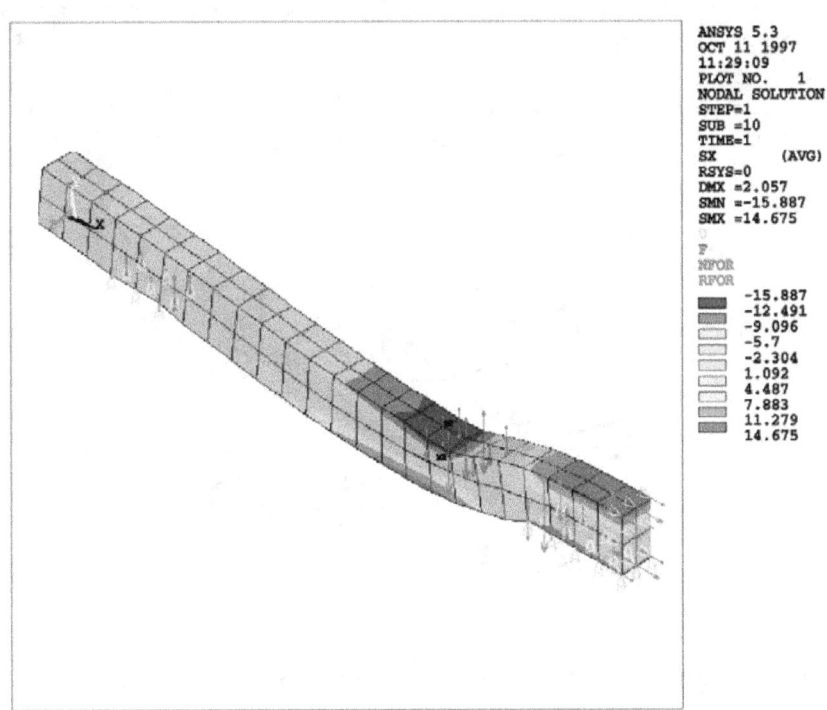

Figure 4.12 - 3 Dimensional ANSYS® Solution to Continuous Beam Example

Table 4.3 - Linear Elastic Orthotropic Comparison of COMAP with ANSYS®

Program	Displacement at 'O' (mm)			Stress at 'O' (MPa)					
	δ_x	δ_y	δ_z	σ_x	σ_y	σ_z	σ_{xy}	σ_{xz}	σ_{yz}
2 - Dimensional COMAP	-0.17	-1.8	-	-14.2	-2.5	-	1.4	-	-
2 - Dimensional ANSYS®	-0.18	-1.8	-	-14.0	-2.7	-	1.3	-	-
3 - Dimensional COMAP	-0.18	-1.9	-0.07	-13.3	-2.9	-0.02	1.2	-0.01	0.47
3 - Dimensional ANSYS®	-0.18	-2.0	-0.06	-14.5	-4.3	-0.4	1.3	-0.02	0.47

4.5.3 Angle Ply Laminate Analyses

The program COMAP is formulated to analyze laminates with varying ply grain angles. To verify this aspect of our program, we use for comparison a program called PC-Laminate, version 1. This program was written by D.W. Radford from TUTE Technology Services. PC-Laminate employs simple mechanics with classical lamination theory (section 4.2.2.3) to calculate linear elastic mechanical responses of laminates with varying ply angles and also provides an estimate for the load to cause the initial failure of the laminate (first ply failure) under in-plane loads using the basic Tsai-Wu criterion (Equation 2.3) with the interaction F_{12} parameter equal to zero.

For the present analysis, $[\pm 15]_s$ and $[\pm 30]_s$ angle-ply laminates were investigated. The stacking sequence of these laminates is illustrated in Figure 4.13 where, in our case, $\theta = 15°$ or $30°$.

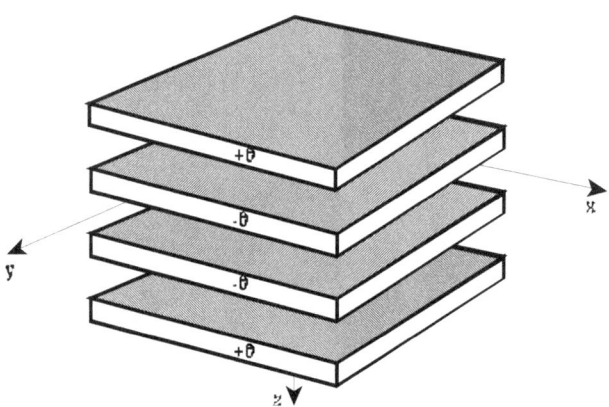

Figure 4.13 - Exploded View of Symmetric Angle-ply Laminate

For both programs, the deterministic elastic and strength properties of the material were input as:

Parallel-to-grain Elastic Modulus : $E_1 = 11$ x 10^3 MPa
Perpendicular-to-grain Elastic Modulus : $E_2 = 0.4$ x 10^3 MPa
Through-thickness Elastic Modulus : $E_3 = 0.62$ x 10^3 MPa (for 3-d analysis only)
Poisson's Ratios: $\nu_{12} = 0.32$; $\nu_{21} = 0.012$;   (for 2-d analysis)
     $\nu_{13} = 0.29$; $\nu_{31} = 0.01$; $\nu_{23} = 0.20$ and $\nu_{32} = 0.20$ (for 3-d analysis only)
Shear Moduli: $G_{12} = 0.7$ x 10^3 MPa   (for 2-d analysis)
     $G_{13} = 0.76$ x 10^3 MPa; $G_{23} = 0.08$ x 10^3 MPa   (for 3-d analysis only)
Parallel-to-grain Tensile Strength: $X_t = 80$ MPa
Perpendicular-to-grain Tensile Strength: $Y_t = 5$ MPa
Parallel-to-grain Compressive Strength: $X_c = 60$ MPa
Perpendicular-to-grain Compressive Strength: $Y_c = 15$ MPa
In-Plane Shear Strength: $S = 6$ MPa
Interaction Parameter: $F_{12} = 0.0$ MPa^{-2}

Both angle ply configurations, $[\pm 15]_s$ and $[\pm 30]_s$, were discretized in the same manner, illustrated in Figure 4.14. In the 2 dimensional model, the plane, 4 node elements were comprised of 4 layers each representing the plys. For each layer, a 2x2 Gauss quadrature rule was used resulting in 16 different stress points per element. The load was applied proportionately at the 5 end nodes, as shown. For the 3 dimensional model, 8 node brick elements were used. The width of each element coincided with the thickness of the ply. The load was spread over the cross section as shown.

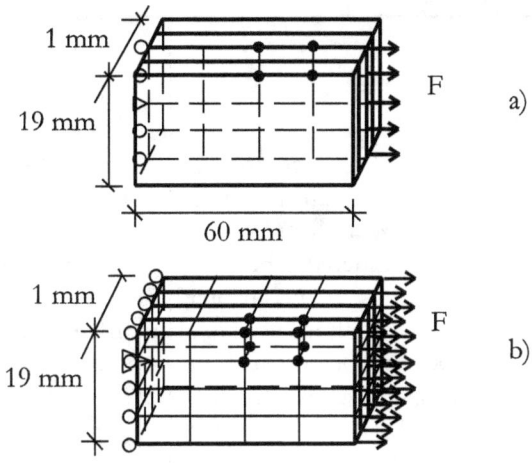

Figure 4.14 - **Finite Element Mesh for Angle Ply Laminates a) 2 dimensions b) 3 dimensions**

Table 4.4 outlines the elastic stiffness and first ply failure results for all analyses. The values agree very well. The only difference worthy of mention is between the first ply failure load result for COMAP (maximum 60.1 MPa) and PC-Laminate (55.9 MPa) with a maximum percentage error of 7%. This is due to the fact that, unlike the result by PC-Laminate, the ultimate stress computed by COMAP includes some stress redistribution prior to failure.

Table 4.4 - Laminate Stiffness and First Ply Failure Comparison of PCLAM and COMAP

Program	$[\pm 15]_s$ Angle - Ply Laminate		$[\pm 30]_s$ Angle - Ply Laminate	
	E_x (MPa)	Ultimate stress, σ_x (MPa)	E_x (MPa)	Ultimate stress, σ_x (MPa)
PCLAM	9080	55.9	4870	23.1
2 - D. COMAP	9083	60.1	4866	23.6
3 - D. COMAP	9072	59.3	4855	23.2

5.1 INTRODUCTION

Evident from the discussion of the yield criterion in Chapter 3, the constitutive properties of the strands (in tension, compression and shear) are required as input to the model. Due to the unique failure characteristics of wood in response to these loadings and with respect to the principal material directions, different methodologies were considered to best estimate each property. Both experimental and analytical approaches were taken. The purpose of this chapter is to both describe fabrication and preparation of test specimens and also delineate the methods used in acquiring the constitutive properties of the strands. As mentioned in the introduction, one potential function of the model developed in this thesis would be to gauge the influence of changing the host material on the final product strength. For example, one could investigate the capacity of a PSL product made from a different wood species. Conceivably, one could perform simple constitutive tests, as outlined in this chapter, on any wood species and use the results as input to the program to obtain an estimate of the properties of the 'new' product.

5.2 MATERIAL

A total of 100 - 3mm x 1220mm x 2440mm sheets of Coastal Douglas-fir veneers were generously supplied by Trus Joist MacMillan Ltd. The veneers were grouped in equal amounts of heartwood and sapwood and were representative of those used in the making of Parallam®, PSL. Veneers chosen for use were deliberately chosen free of large defects such as knots or localized slope of grain. All specimens described in the following sections were prepared and fabricated by the author from these veneers.

5.3 EXPERIMENTAL MEASUREMENT OF STRAND PROPERTIES

Standard test methods have been established for testing of wood in tension and compression by the American Society for Testing and Materials (ASTM). Although no specific tests are identified by these standards for testing of individual wood strands (specimens were nominally 19mm in width to emulate that used in PSL), ASTM standards were used as general guidelines.

Specimens were conditioned to the ambient laboratory environment as opposed to being conditioned in a controlled humidity chamber due to a lack of this equipment at one of the test facilities. All compression specimens were tested in ambient environmental conditions of average 21.9°C ± 1.3°C and 30.2% ± 5.8% relative humidity. This correlates to an average of approximately 6.3% ± 1% moisture content (Siau, 1984). For all other cases, specimen moisture content was monitored by use of the oven dry method (ASTM D2016) (mean = 7.7%, standard deviation = 0.9%).

5.3.1 Compression Tests

5.3.1.1 Material Preparation

To overcome practical difficulties with testing equipment when dealing with the potential of buckling of thin specimens (such as working with an unpractically short gauge length to satisfy short column requirements), the compression tests were performed on 6-ply laminated specimens. These specimens (as well as specimens for other configurations discussed in the next sections) were cut from boards which were fabricated in the laboratory as follows:

3 x 610 x 610 mm^2 Douglas-fir heartwood veneers were first conditioned to a moisture content of approximately 4 percent to avoid blowouts. Then, a phenol-formaldehyde resin (PF 355H - a thermosetting resin used in making commercial wood composites) was combined in accordance with a standard procedure with pure dried cob, flour, soda ash and water, and then applied to the veneers using a roll spreader. The roll spreader provided for a consistent method of applying a prescribed amount of resin to each veneer. Six sheets of veneer oriented with the face grain in the longitudinal direction (such that the lathe checks for the bottom three veneers faced and opposed that for the top three veneers to avoid cupping) were laminated and pressed using a 1600 KN capacity Pathex 760 x 760 mm^2 hot-press at 150^{0}C and 1.38 MPa pressure for 6 minutes. The same procedure was done using Douglas-fir softwood veneers.

Compression 'parallel-to-grain' specimens (nominally 17 x 19 x 60 mm^3) and compression 'perpendicular-to-grain' specimens (17 x 19 x 50 mm^3) were cut from 8 separate boards - 4 boards of heartwood and 4 of softwood. The specimens were cut using extreme care to ensure right angles to minimize eccentricities in loading. Furthermore, in the center of the cross-section of each specimen, a small divot (diameter less than 1 mm) was drilled to align the specimen with the center of the load platens which, in turn, had a matching small point. Parallel-to-grain specimen length was determined in accordance with ASTM D198 for short columns with no lateral support. This requirement states that the length to least radius of gyration not exceed 17.

5.3.1.2 Test Method

Compression tests were conducted at the Laboratoire de Mechanique et Technologie (LMT), École Normale Supérieure (ENS), Cachan - University Paris, France. Tests were performed on a servo- hydraulic Material Testing System (MTS) machine with capacity 50KN under deflection control mode. The setup was equipped with a Linear Variable Displacement Transducer (LVDT) to measure cross head displacement. Swivel bearings, both top and bottom, were used to prevent eccentric loading on the specimen as shown in Figure 5.1. An MTS 632.c20 extensometer which is typically used with smooth surface advanced composites, was employed for some tests but gave inconsistent results due to poor contact with the wood surface. This approach was therefore abandoned. Instead, elastic moduli were determined using the cross-head displacement. To remove error caused by the inherent flexure in the testing machine, a compression test was performed with steel plates only. The resulting

Figure 5.1 - Compression Test Set-up

displacement, attributable to the machine, was thereby deducted from the overall displacement in the calculation of the material stiffness. Machine displacement was found to account for up to 30 percent of total displacement.

Cross sectional dimensions were measured for each specimen using calipers with an accuracy of ±0.005 mm. A uniform loading rate of 0.12 mm/min. was used to produce failure between 5 and 10 minutes in compliance with ASTM D198.

5.3.1.3   Experimental Results

Typical stress-strain curves for the parallel-to-grain and perpendicular-to-grain specimens are given in Figures 5.2 and 5.3, respectively. The curves have been zero-adjusted to remove nonlinearites at the curve origin. This was done by shifting the curve in the x direction only, such that the projection of the linear portion of the stress-strain curve passes through the origin. These nonlinearities occur due to the settling of the specimen within the rotating spherical bearing blocks. Beyond the initial linear portion of the curve, however, there is a notably nonlinear material response, as is expected for wood in compression (Goodman and Bodig 1971; Maghsood et al. 1973).

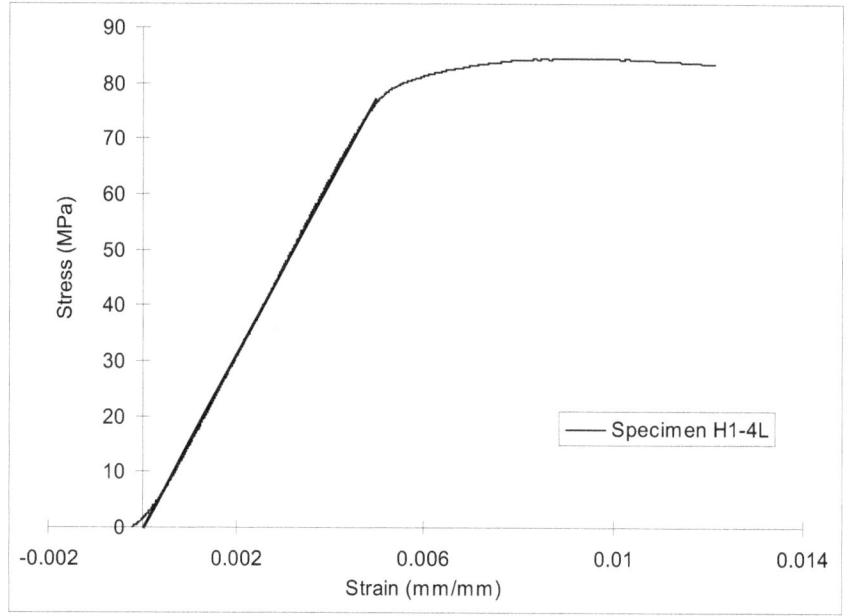

Figure 5.2 -   Typical Stress-Strain Curve for Compression Parallel-to-Grain Specimens

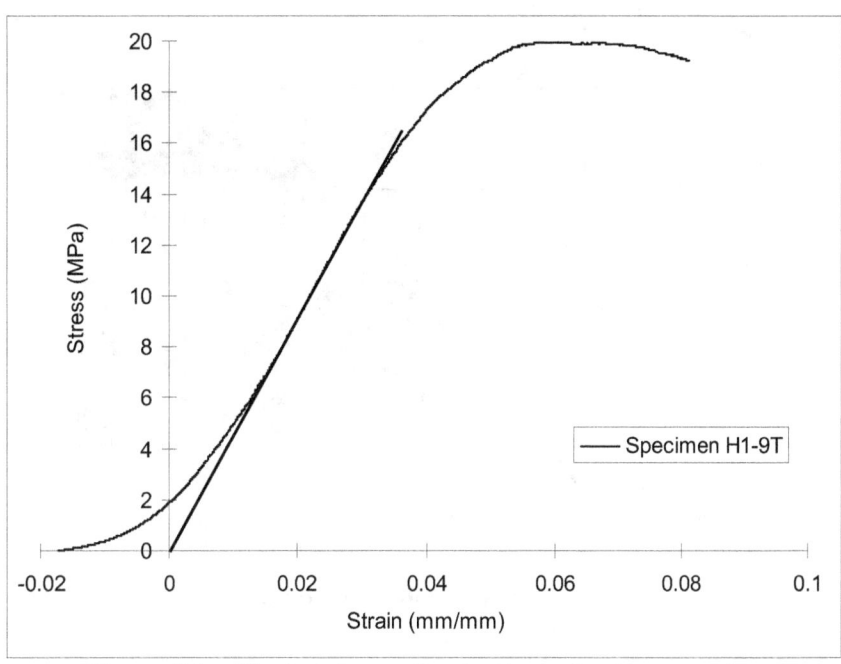

Figure 5.3 - Typical Stress-Strain Curve for Compression Perpendicular-to-Grain Specimens

In order to describe this mechanical behaviour for use in the computer model, the curve is simplified to a tri-linear approximation with 4 defining variables: the elastic modulus prior to yielding, the yield stress, the elasto-plastic tangent modulus beyond yield and the ultimate stress (denoted for a perpendicular-to-grain specimen: E_{2c}, Y_c, E_{2c}', and Y_c^u, respectively). These values were determined for each specimen by equating the area under the experimental stress-strain curve (*i.e.* the strain energy stored in the specimen) to the area of the fitted tri-linear stress-strain curve as illustrated in Figure 5.4. Perfect plasticity beyond ultimate load (as opposed to strain softening) has been assumed. This assumption not only simplifies the analysis greatly, but also agrees with experimental results up to an acceptable strain level.

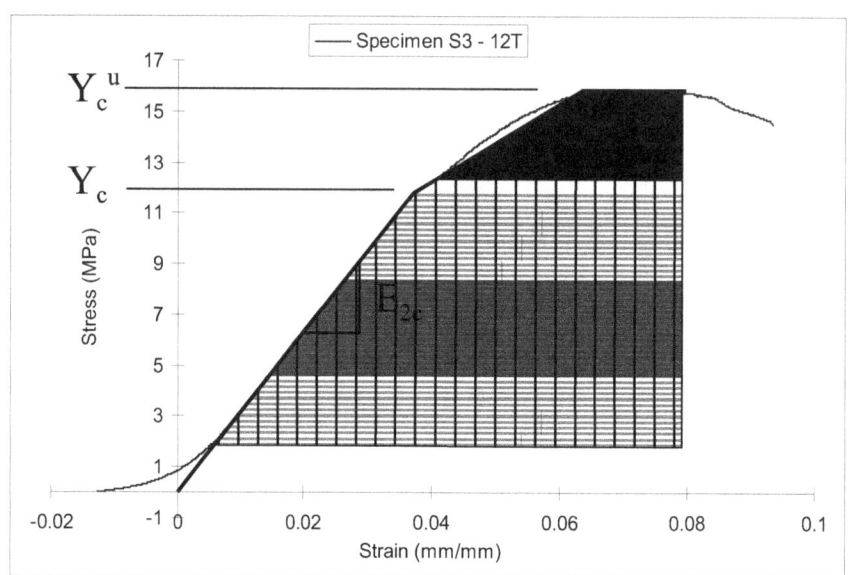

Figure 5.4 - Tri-linear Approximation of Compression Stress-Strain Curve

A comprehensive summary of all measured compressive properties is given in Table 5.1. Descriptive statistics are provided for each set of data.

Table 5.1 - Compression Properties of Strands

Configuration		Count	Property							
			Elastic Modulus		Yield Stress		Tangent Modulus		Ultimate Stress	
			Mean (GPa)	COV. (%)	Mean (MPa)	COV. (%)	Mean (GPa)	COV. (%)	Mean (MPa)	COV. (%)
Heartwood	Parallel	53	10.09	19.13	67.32	18.04	1.93	33.11	76.46	7.06
	Perp.	54	0.49	15.23	15.37	11.75	0.11	35.05	18.19	9.28
Sapwood	Parallel	51	11.72	27.65	65.16	10.14	2.47	36.56	73.61	9.64
	Perp.	56	0.37	13.54	14.75	20.68	0.10	33.13	16.71	18.09

A single factor Analysis of Variance (ANOVA) was performed for the heartwood and sapwood results for each property. The results of this analysis is outlined in Tables 5.2 and 5.3. The two data sets broken down by descriptive property elastic modulus, yield stress, etc., for both parallel-to-grain and perpendicular-to-grain samples, were found to be significantly different at a 0.05 percent level of significance. Consequently, the two sets of data could not be combined. It was decided arbitrarily that further development of the strand database and thus the model, would focus solely on the heartwood material.

Table 5.2 - Analysis of Variance Results for Heartwood and Sapwood for Parallel-to-grain Compression

Dependent Variable	Source of Variation	Degree of Freedom	Sum of Squares	Mean Square	F Value	$F_{0.05}$
Elastic Modulus	Between groups	1	14434405.5	14434405.5	2.1	3.9
	Within groups	97	676023834.2	6969317.9		
	Total	98	690458239.6			
Yield Stress	Between groups	1	116.1	116.1	3.2	3.9
	Within groups	97	3534.4	36.4		
	Total	98	3651.5			
Tangent Modulus	Between groups	1	7228897.3	7228897.3	11.9[1]	3.9
	Within groups	97	58960353	607838.7		
	Total	98	66189250			
Ultimate Stress	Between groups	1	184.3	184.3	4.6[1]	3.9
	Within groups	97	3858.3	39.8		
	Total	98	4042.7			

[1] significantly different at a 0.05 percent level of significance

Table 5.3 - Analysis of Variance Results for Heartwood and Sapwood for Perp.-to-grain Compression

Dependent Variable	Source of Variation	Degree of Freedom	Sum of Squares	Mean Square	F Value	$F_{0.05}$
Elastic Modulus	Between groups	1	330105.8	330105.8	79.5[1]	3.9
	Within groups	100	415164.6	4151.6		
	Total	101	745270.4			
Yield Stress	Between groups	1	9.5	9.5	1.6	3.9
	Within groups	100	607.5	6.1		
	Total	101	617.0			
Tangent Modulus	Between groups	1	19.1	19.1	0.02	3.9
	Within groups	100	127398.8	1274.0		
	Total	101	127418.0			
Ultimate Stress	Between groups	1	51.0	51.0	8.8[1]	3.9
	Within groups	101	577.4	5.8		
	Total	102	628.3			

[1] significantly different at a 0.05 percent level of significance

The nonlinear compressive behaviour of the strands is randomly generated within the model based on the descriptive statistics of the four defining variables which are appropriately correlated. Each variable is simulated according to a bivariate standard normal distribution (Wang and Lam, 1998):

$$b_i = \mu_{bi} + \sum_{j=1}^{4} L_{ij}z_j \quad (i = 1,2,3,4) \tag{5.1}$$

where: (for parallel to grain properties, for example) $b_1 = E_{1C}$, $b_2 = X_c$, $b_3 = E_{1C}'$, $b_4 = X_c^u$, μ_{bi} is the respective variable mean, L_{ij} is the lower triangle of the correlation matrix of the variables, derived through Cholesky decomposition, and z_i is an independent standard normal random variable. The correlation matrix and corresponding lower triangle matrix are given for both parallel-to-grain and perpendicular-to-grain properties in Tables 5.4 through 5.7.

Table 5.4 - Correlation Matrix for Compression Parallel-to-Grain Specimens

Property	Elastic Modulus	Yield Stress	Tangent Modulus	Ultimate Stress
Elastic Modulus	1.00	0.48	0.20	0.55
Yield Stress	0.48	1.00	-0.21	0.84
Tangent Modulus	0.20	-0.21	1.00	0.14
Ultimate Stress	0.55	0.84	0.14	1.00

Table 5.5 - Lower Triangle of Correlation Matrix for Compression Parallel-to-Grain Specimens

Property	Elastic Modulus	Yield Stress	Tangent Modulus	Ultimate Stress
Elastic Modulus	1.00	0.00	0.00	0.00
Yield Stress	0.48	0.88	0.00	0.00
Tangent Modulus	0.20	-0.35	0.91	0.00
Ultimate Stress	0.55	0.65	0.28	0.43

Table 5.6 - Correlation Matrix for Compression Perpendicular-to-Grain Specimens

Property	Elastic Modulus	Yield Stress	Tangent Modulus	Ultimate Stress
Elastic Modulus	1.00	0.13	0.10	0.37
Yield Stress	0.13	1.00	-0.42	0.81
Tangent Modulus	0.10	-0.42	1.00	0.07
Ultimate Stress	0.37	0.81	0.07	1.00

Table 5.7 - Lower Triangle of Correlation Matrix for Compression Perpendicular-to-Grain Specimens

Property	Elastic Modulus	Yield Stress	Tangent Modulus	Ultimate Stress
Elastic Modulus	1.00	0.00	0.00	0.00
Yield Stress	0.13	0.99	0.00	0.00
Tangent Modulus	0.11	-0.43	0.90	0.00
Ultimate Stress	0.37	0.77	0.40	0.34

Figures 5.5 through 5.12 show the experimental data points in terms of cumulative probability distributions of the four defining variables together with fitted bivariate normal distributions. Through visual inspection, the experimental results are very well represented by the simulated data.

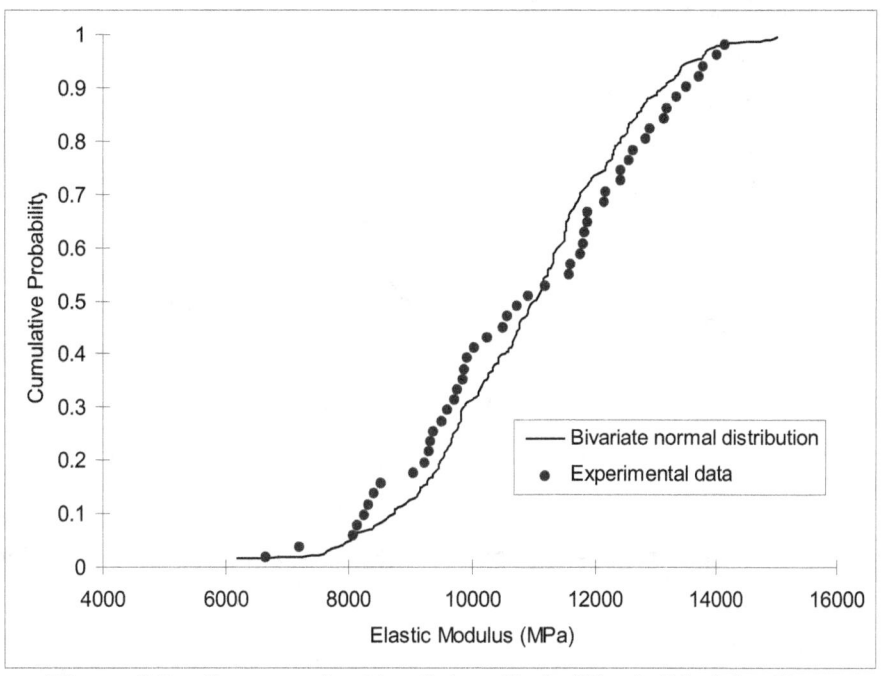

Figure 5.5 - Compression Parallel-to-Grain Elastic Modulus Data

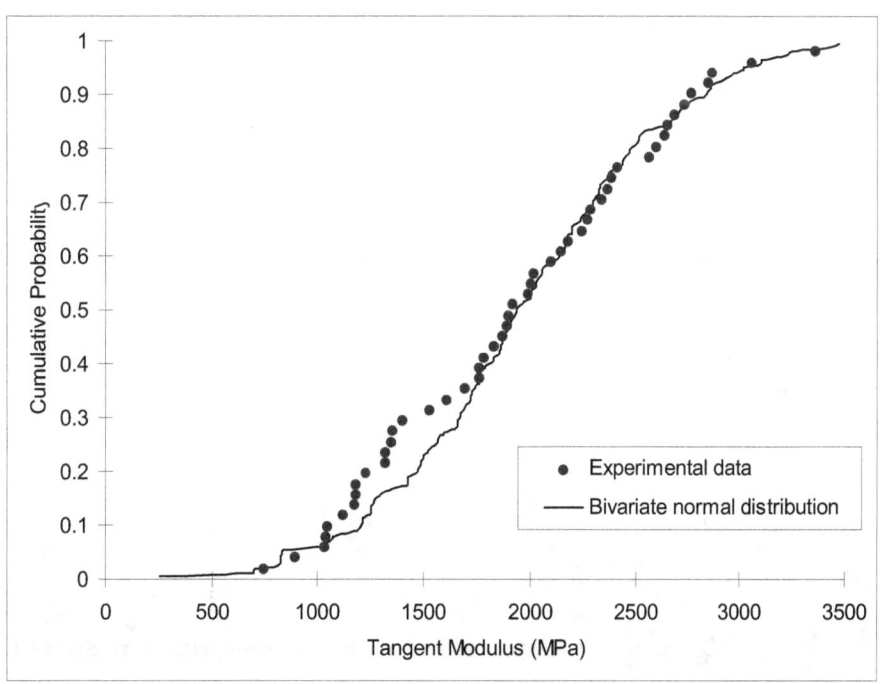

Figure 5.6 - Compression Parallel-to-Grain Tangent Modulus Data

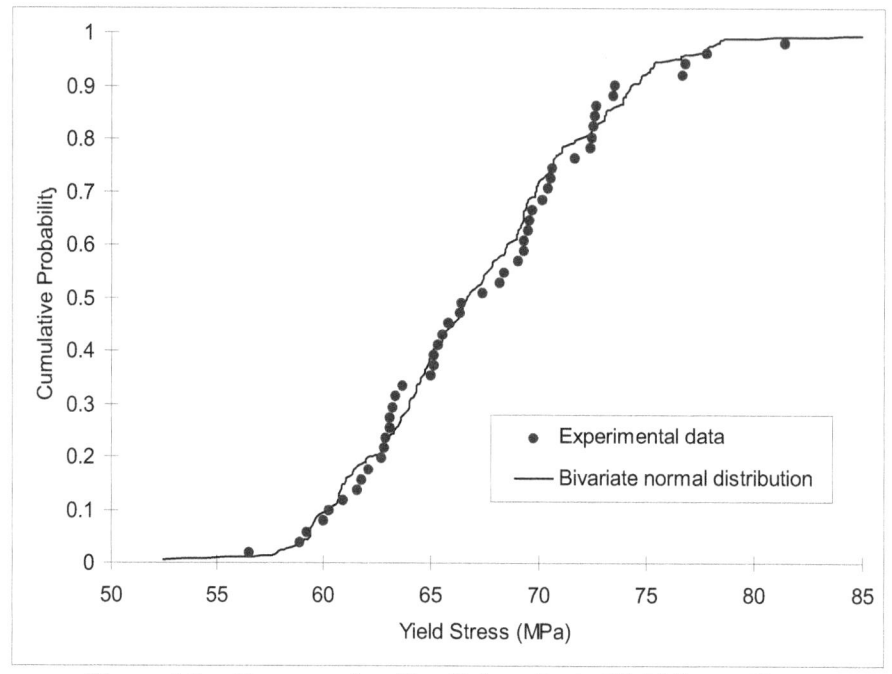

Figure 5.7 - Compression Parallel-to-Grain Yield Stress Data

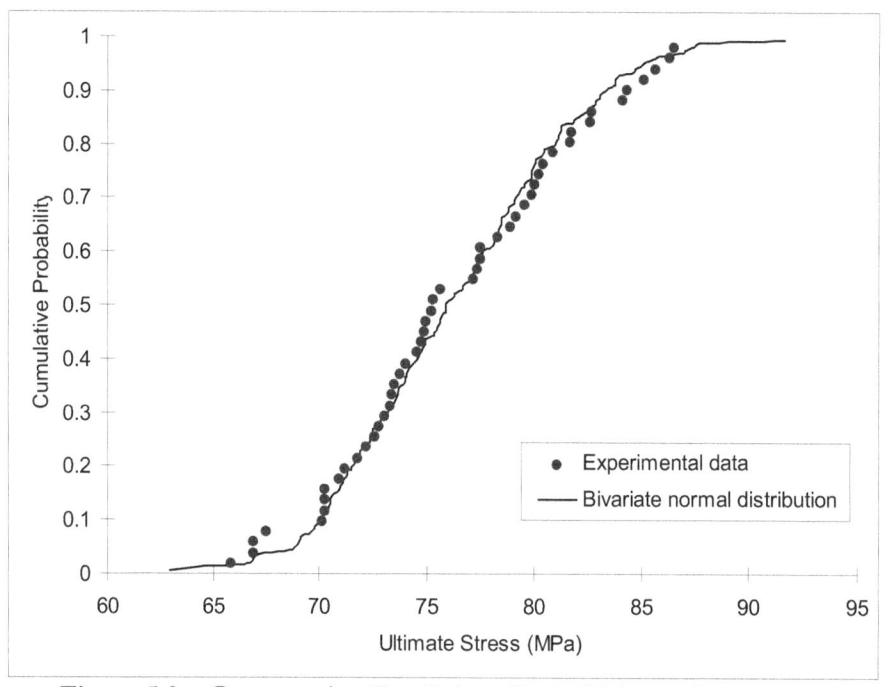

Figure 5.8 - Compression Parallel-to-Grain Ultimate Stress Data

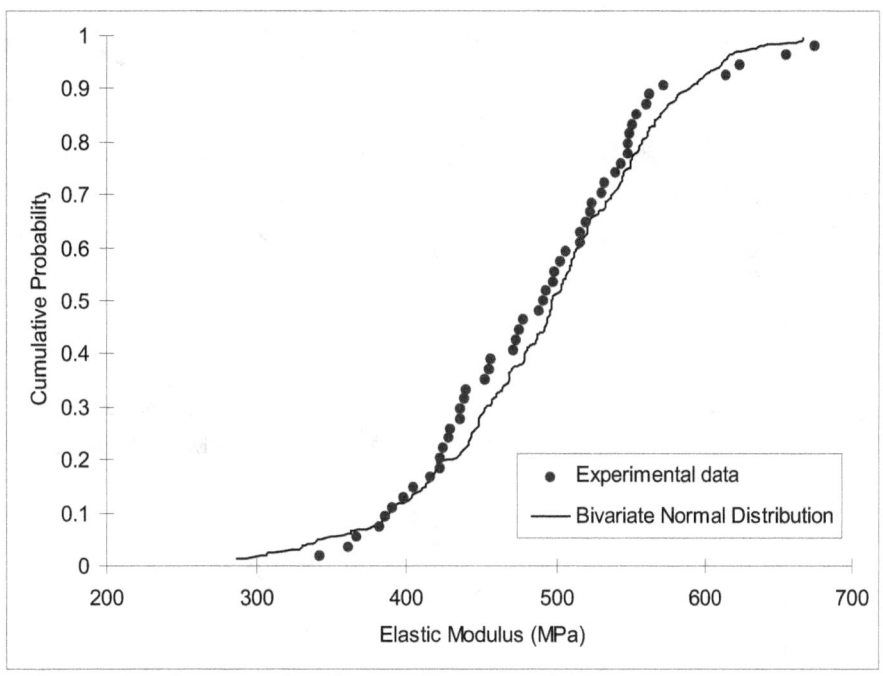

Figure 5.9 - Compression Perpendicular-to-Grain Elastic Modulus Data

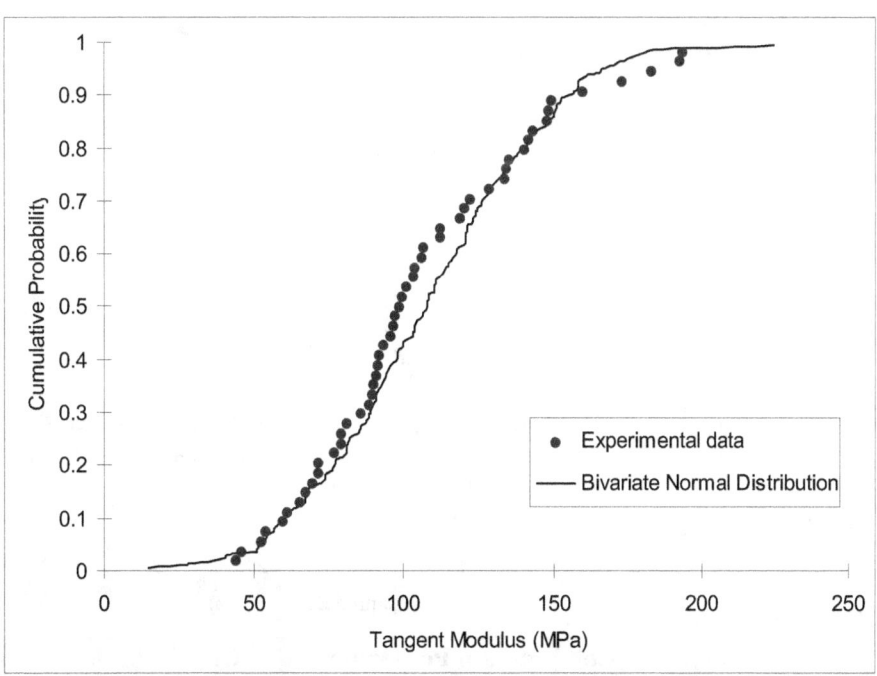

Figure 5.10 - Compression Perpendicular-to-Grain Tangent Modulus Data

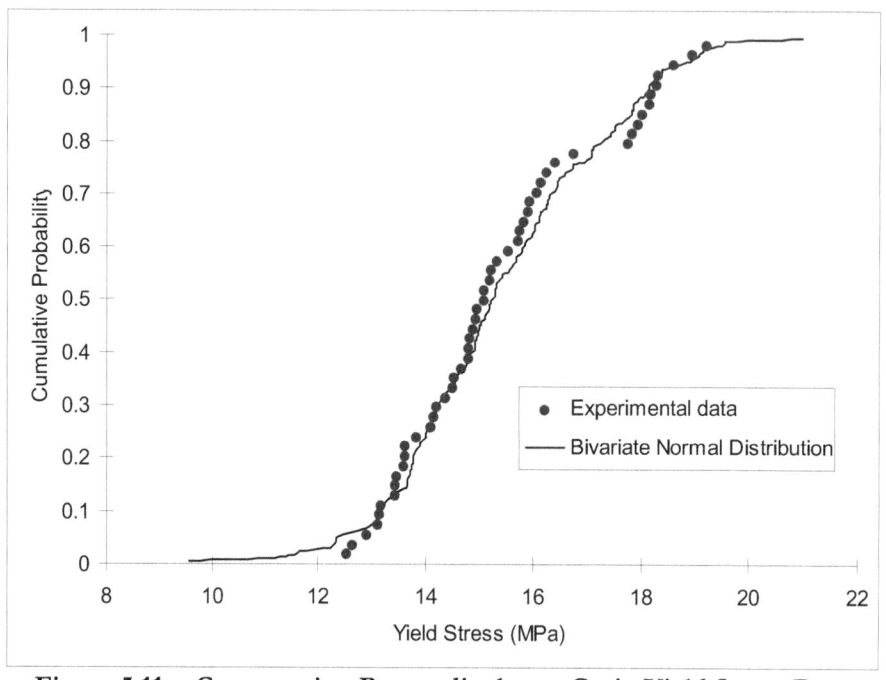

Figure 5.11 - Compression Perpendicular-to-Grain Yield Stress Data

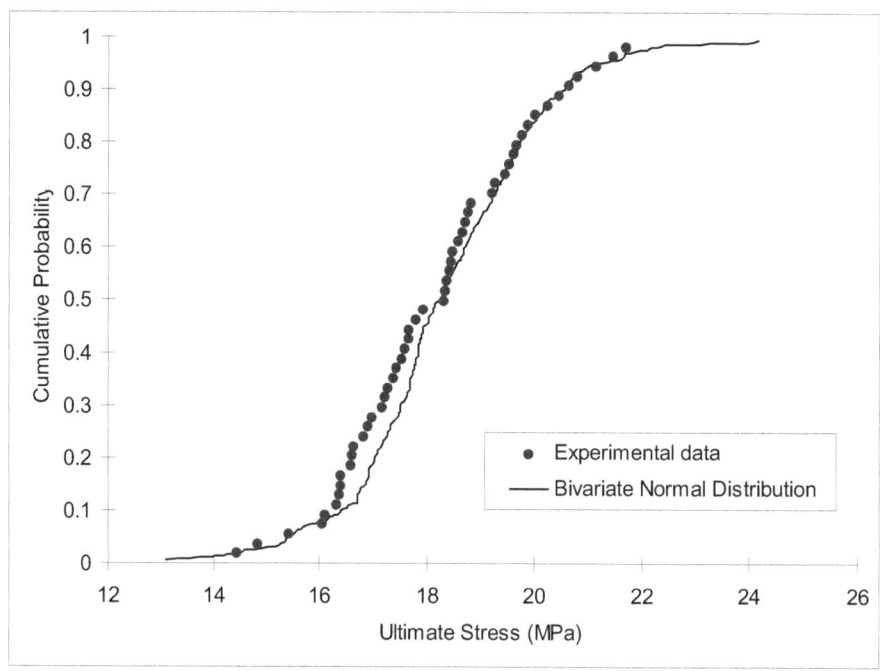

Figure 5.12 - Compression Perpendicular-to-Grain Ultimate Stress Data

5.3.2 Tension tests

5.3.2.1 Material Preparation

For parallel-to-grain tests, individual strands (3 x 19 x 50 mm^3) were cut from 8 separate sheets of Douglas-fir heartwood veneer in the longitudinal orientation. For perpendicular to grain tests, in order to avoid breakage during handling, specimens were cut from 6-ply laminated veneer boards (fabricated as described in section 5.3.1). These specimens (17 x 19 x 152 mm^3) were sawn from 4 separate boards using a fine carbide-tipped cut-off saw blade to minimize chipping.

5.3.2.2 Test Method

All tension tests were performed at the University of British Columbia on a 250KN MTS machine together with MTS mechanical wedge action grips and an extensometer for calculation of elastic modulus, as depicted in Figure 5.13.

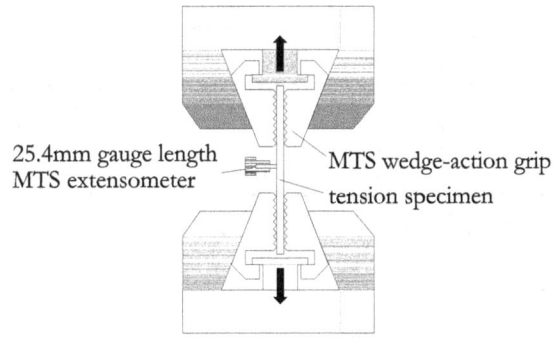

25.4mm gauge length
MTS extensometer

MTS wedge-action grip

tension specimen

Figure 5.13 - Tension Test Set-up

Cross sectional dimensions were measured using calipers. The loading rate was set at 0.64 mm/minute and to produce failure in no less than 5 minutes and no greater than 10 minutes in compliance with ASTM D198. Ultimate load and description of failure were recorded. Although rare, specimens which failed in the grip region were discarded.

5.3.2.3 Experimental Results

A comprehensive summary of all measured mechanical properties is given in Table 5.8. Typical stress-strain curves for the parallel-to-grain and perpendicular-to-grain specimens are given in Figures 5.14 and 5.15, respectively. The strain in these graphs is calculated using crosshead movement divided by gauge length. It is noted, however, that the crosshead movement includes inherent grip movement. Referencing Figure 5.13, as the load increases, the specimen crushes around the grip teeth, which precipitates vertical displacement within the wedge action grips. This occurs prominently at initial load (leading to nonlinearities), and continues throughout loading to a lesser degree as the specimen becomes more dense. Consequently, the results have been zero-adjusted to remove these nonlinearities at the curve origin per section 5.3.1.3. Also, as a result, the graphs appear more compliant and do not represent a true account of tensile stiffness. Tensile stiffness, instead, was determined through use of the extensometer. The extensometer measures only local displacement and must be removed prior to ultimate failure of the specimen to avoid instrument damage. As such, it was not an alternative to calculate strain in Figures 5.14 and 5.15.

The graphs demonstrate the linear elastic behaviour of wood when loaded in tension. In view of this, two variables only were used to define and regenerate the tensile behaviour in each direction for the model: E_{1t}, X_t, and E_{2t}, Y_t - elastic modulus and ultimate stress in the parallel and perpendicular direction, respectively. Elastic modulus was determined for each specimen through linear regression of the stress-strain curve where strain was measured using the 25.4 mm gauge extensometer.

Figures 5.16 through 5.19 show the data points in terms of cumulative probability distributions of elastic modulus and ultimate stress together with fitted log-normal distributions.

Table 5.8 - Tension Properties of Strands

Configuration	Count	Property			
		Elastic Modulus		Ultimate Stress	
		Mean (GPa)	COV (%)	Mean (MPa)	COV (%)
Parallel-to-grain	36	15.46	30.50	68.77	26.68
Perpendicular-to-grain	45	0.09	24.40	1.91	18.31

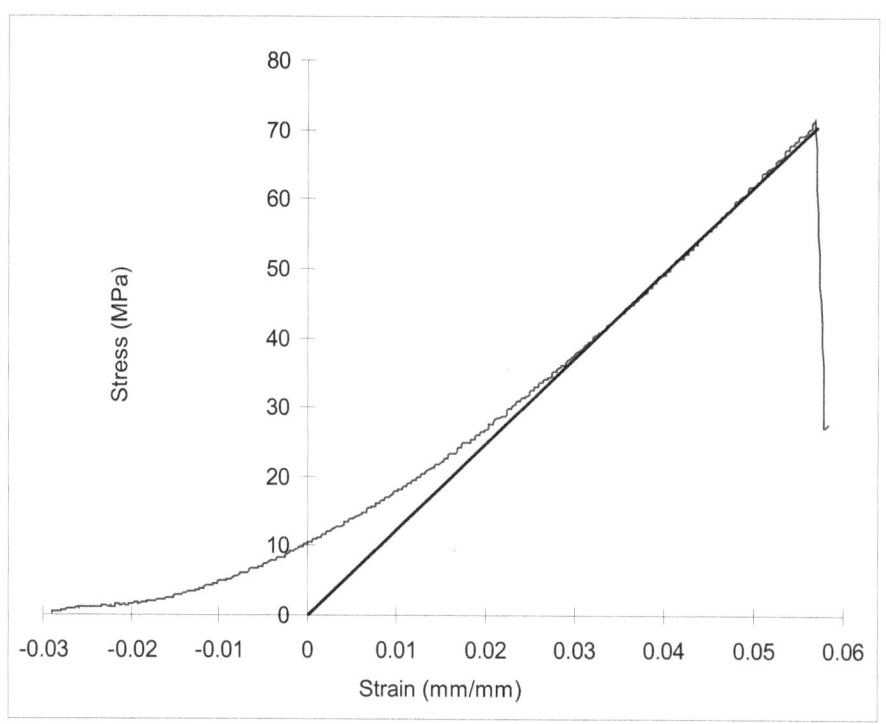

Figure 5.14 - Typical Parallel-to-Grain Tensile Behaviour

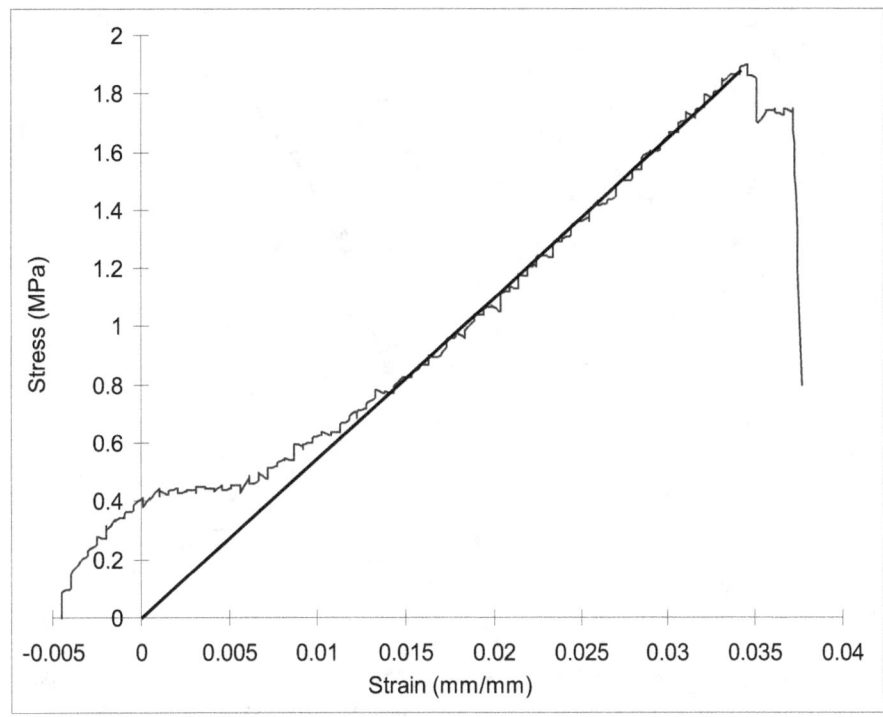

Figure 5.15 - Typical Perpendicular-to-Grain Tensile Behaviour

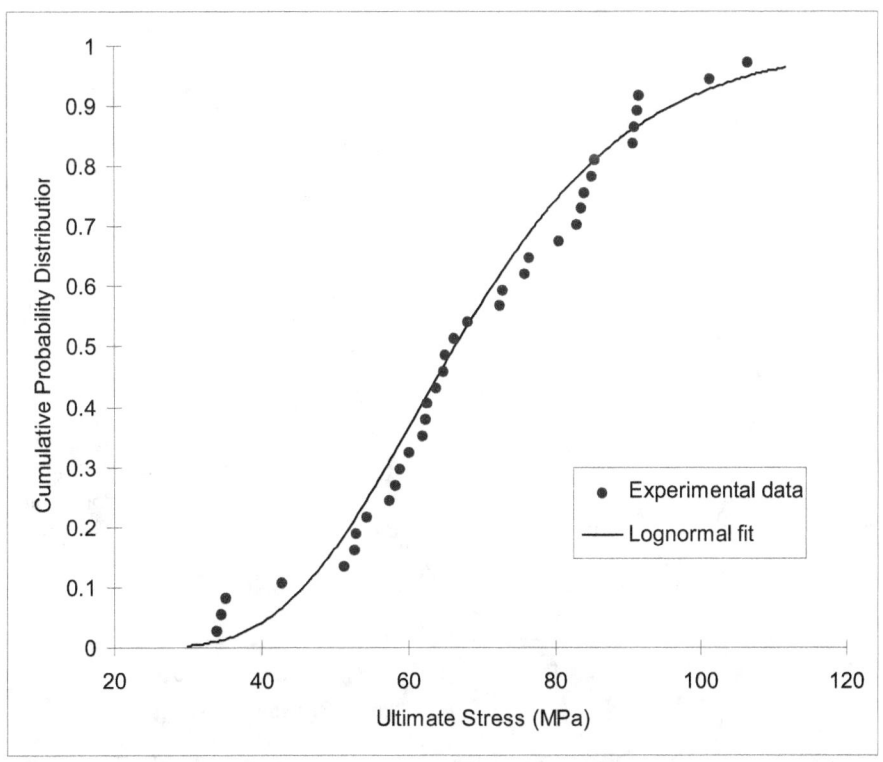

Figure 5.16 - Lognormal fit of Parallel-to-Grain Tensile Strength Data

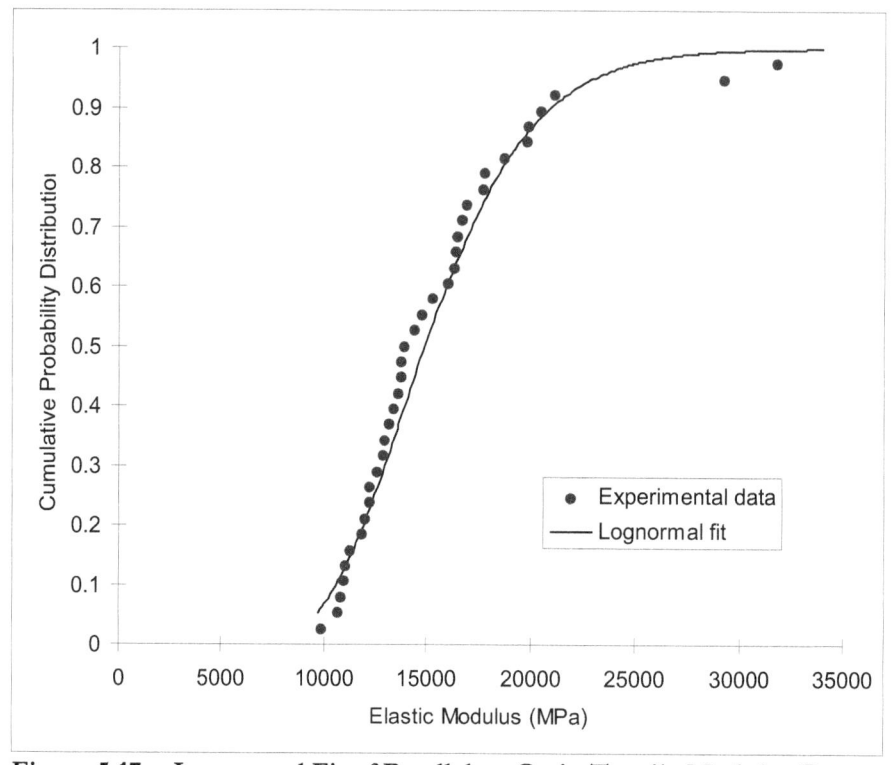

Figure 5.17 - Lognormal Fit of Parallel-to-Grain Tensile Modulus Data

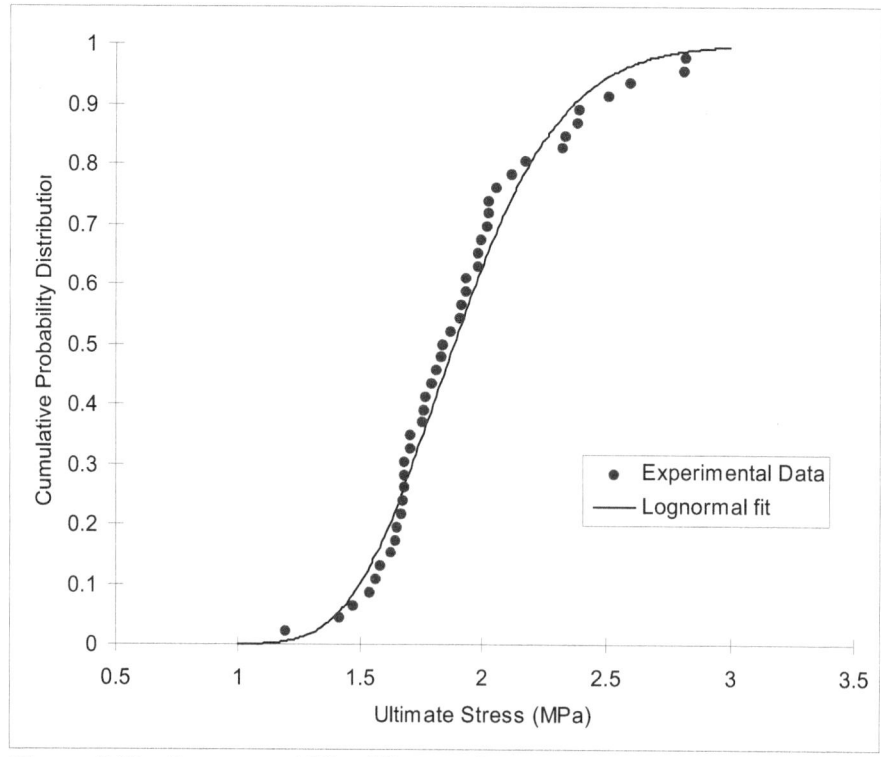

Figure 5.18 - Lognormal Fit of Perpendicular-to-Grain Tensile Strength Data

The tensile strengths (both parallel and perpendicular-to-grain) reflect only the strength of the corresponding tested volume. Prior to implementing the values into the finite element code, adjustments must first be made for size.

5.3.3 Strength Modification due to Size Effect

It has long been recognized for brittle materials that large members tend to display lower strengths than smaller members when subjected to the same environmental and loading conditions. The phenomenon has come to be known as 'size effect' and has implications on the tensile strength data discussed in the previous section. Numerous researchers have acknowledged and addressed size effect for many different materials (Barrett et al., 1975; Madsen, 1990; Sharp and Suddarth, 1991; Zweben, 1994). Each of these studies espoused the use of Weibull weakest-link theory to quantify size effect. As such, this theory was adopted for this study.

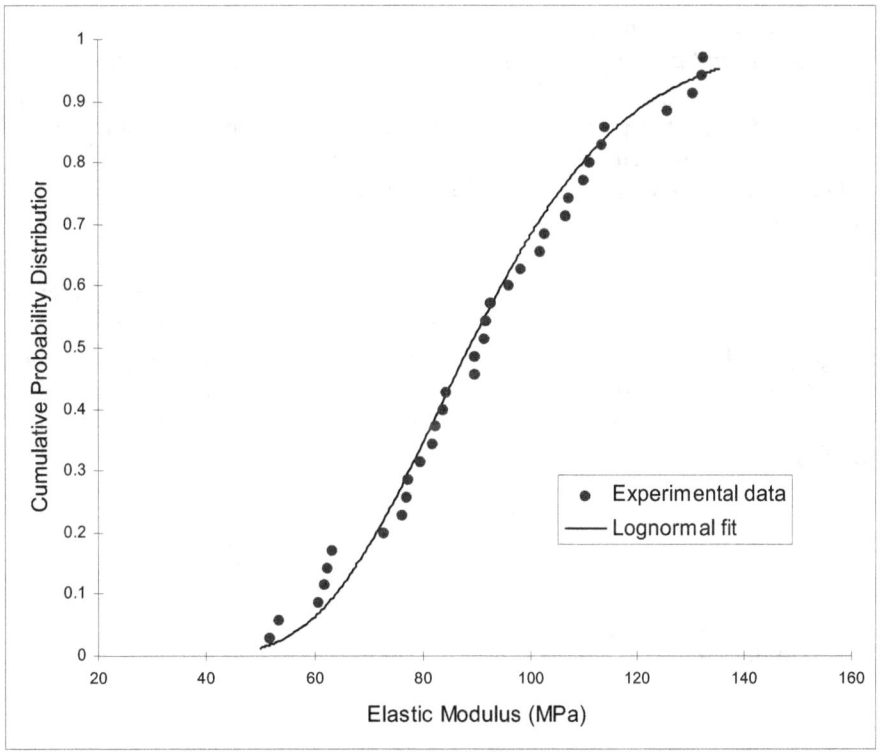

Figure 5.19 - Lognormal Fit of Perpendicular-to-Grain Tensile Modulus
                Data

5.3.3.1 Weibull Weakest-Link Theory

Weibull (1939) proposed that a variation in strength properties existed due to the statistical probability of a strength controlling defect occurring in a given volume. He used the weakest link concept to show how strength of a 'perfectly brittle' material could be described by a specific cumulative distribution function. Weibull's theory enabled the prediction of the probability of failure of a homogeneous isotropic material at a given volume according to the following:

$$F(\tau) = 1 - e^{-\frac{1}{V_o} \int_V \left[\frac{\tau - \tau_{min}}{m} \right]^{\beta} dV} \tag{5.2}$$

where :

$F(\tau)$ = probability of failure
τ = material strength
τ_{min} = minimum material strength (location parameter)
m = scale parameter
β = shape parameter
(τ_{min}, m, and β are material constants and V_o is a reference volume)

Commonly, τ_{min} is accepted as being equal to zero, simplifying the problem, and the resulting formulation is called a two-parameter Weibull distribution. This distribution may be used to explain the strength difference observed in material bodies subjected to the same stress distribution but differing in volume as follows:

It is argued that a larger member has a higher probability of containing a larger flaw (or weaker zone) than does the smaller member and thus has a lower strength. In general, the strength of a volume of material at a given probability of failure is predicted given the strength and shape parameter of a common material when both are subjected to the same stress distribution. The solution is found by equating the probability of failure of one volume, V_1 (corresponding to τ_1), to that of another volume, V_2 (corresponding to τ_2), which yields:

$$1 - e^{-\frac{1}{V_o} \int_{V_1} \left[\frac{\tau_1}{m} \right]^{\beta} dV_1} = 1 - e^{-\frac{1}{V_o} \int_{V_2} \left[\frac{\tau_2}{m} \right]^{\beta} dV_2} \tag{5.3}$$

This simplifies to

$$\int_{V_1} \tau_1^{\beta} dV_1 = \int_{V_2} \tau_2^{\beta} dV_2 \tag{5.4}$$

If it may then be assumed that the stresses in both volumes are uniform throughout the cross sections, then Equation 5.4 becomes:

$$\tau_1^{\beta} V_1 = \tau_2^{\beta} V_2 \tag{5.5}$$

Or, given that the cross section dimensions are equal:

$$\tau_1^{\beta} L_1 = \tau_2^{\beta} L_2 \tag{5.6}$$

As illustrated in Figures 5.14 and 5.15, wood displays brittle behaviour under tensile loading. Based on this, a size effect was assumed to exist and consequentially, the average values of the tensile strengths (both parallel and perpendicular-to-grain) were adjusted from the test volume to the representative size in the finite element analysis according to either Equation 5.5 or 5.6.

5.4 ANALYTICALLY DERIVED PROPERTIES

Various approaches have been taken in the past for determining shear strength and stiffness of wood - the more popular being the ASTM shear block method for small clear specimens described in ASTM D143. This method, however, has limitations in that the test setup creates a complex distribution of stresses in the specimen and hence a questionable estimation of pure shear strength. Also, an added challenge lays in finding the pure shear characteristics of a wood strand (ie. from practical difficulties in testing the strand due to its geometry). At the same time there are complications, as outlined in Section 2.2.1.2, with establishing the interaction parameter of the Tsai-Wu criterion. An alternate method to obtain all three required variables (shear strength, S; modulus of rigidity, G; and the interaction parameter, F_{12}), based loosely on the unidirectional polymer matrix composite standard, ASTM D3518 / D3518M, has been developed for this study.

This standard employs a $[\pm45]_s$ angle-ply laminate subjected to uniaxial tension. It can be shown that the lamina shear stress (σ_{12}) is related to the uniaxial tensile stress on the laminate (σ_x) by

$$\left|\sigma_{12}\right| = \left|\frac{\sigma_x}{2}\right| \tag{5.7}$$

and hence an estimate of in-plane shear strength (S) can be calculated from the ultimate tensile stress. Although the standard advocates this method for shear strength estimation, it is noted that the lamina is still not in a state of pure shear, as normal stress components are present in the transformed coordinate system. As there is a known multiaxial state of stress present, it would be more logical to employ a strength criterion. Accordingly, the following approach has been developed to estimate the mean and standard deviation of the shear properties and the interaction parameter of wood strands.

5.4.1 Minimization Technique

The statistical parameters of the variables were approximated simultaneously following a minimization technique originally developed for the Foschi-Yao damage accumulation model (Foschi.and Yao, 1986). The procedure entails a nonlinear least square minimization of error (using a quasi-Newton approach) between predicted and experimental compression strengths of a $[\pm15]_s$ angle-ply laminate. The compressive strengths of the laminates were simulated using the finite element model described in Chapter 4 together with Monte Carlo simulations. The predicted laminate strength is a function of the mean and standard deviation of the three unknown variables (S, G, F_{12}). Thus, a vector of six unknowns in total $\Lambda = \left\{\mu_S, \mu_G, \mu_{F_{12}}, SD_S, SD_G, SD_{F_{12}}\right\}^T$ is solved in the minimization process. An outline of the procedure follows:

1. Initial estimates for the mean and standard deviation of the three variables were provided for the model. A normal distribution was deemed appropriate for F_{12}, as this enables either positive or negative values, reflecting this characteristic of the parameter. Lognormal distributions were chosen to represent the shear properties.

2. Strength was calculated for a sample of 300 replications. A larger sample could also have been used; however, the number of replications was needed to address computer time restrictions. For each replication the entire load displacement behaviour was modelled in order to establish ultimate stress. For the three dimensional model, each replication took between 20 to 30 minutes using a MS Windows [®] based, 233 MHz, Pentium II system.

3. These predicted strengths were ranked (ie. sorted in ascending order and given the associated probability of failure).

4. A residual function Φ was calculated as:

$$\Phi = \sum_{i=1}^{P} \left(1 - \frac{\tau_i^{pred}}{\tau_i^{exp}} \right)^2 \tag{5.8}$$

where: P denotes the number of probability levels for consideration and the superscripts *pred* and *exp* refer to the predicted and experimental laminate strengths, respectively. The predicted strength was determined for the same i[th] probability level as the experimental strength.

5. As required by the quasi-Newton method, the gradient of the residual function with respect to the

unknowns $\quad \nabla\Phi = \left\{ \dfrac{\partial\Phi}{\partial\Lambda_1}, ... \dfrac{\partial\Phi}{\partial\Lambda_6} \right\}^T \quad$ was calculated. This was done numerically, based on a

perturbation process as follows: each unknown Λ_i was perturbed by a positive increment $0.001\Lambda_i$. Then Φ was recalculated, designated Φ^+. The same was done for the same negative increment producing Φ^-. The gradient was then calculated as:

$$\frac{\partial\Phi}{\partial\Lambda_i} = \frac{\Phi^+ - \Phi^-}{2 \cdot \Delta\Lambda_i} \quad (i = 1,...6) \tag{5.9}$$

6. New, adjusted, statistical parameters replaced the initial estimated values and the residual function was re-evaluated. This procedure was repeated until the difference between residual function values for subsequent iterations satisfied a set tolerance of 1.0×10^{-3}.

It is noted that the $[\pm 15]_s$ angle-ply laminate was determined to be the most suitable configuration for determining both shear and the interaction parameter together. The reason stems from a finding in Clouston et al. (1998). Although the 45^0 configuration would produce the largest in-plane shear stresses (which would be beneficial in estimating shear strength), it was shown through a sensitivity analysis that the data from 15^0 off-axis tests had more tolerance for inaccuracies in experimental strength results and consequently were more reliable than for other angles (30^0, 45^0, or 60^0) for calculating the interaction parameter, F_{12}.

5.4.2 Experimental Tests

5.4.2.1 Material Preparation

Inherent in the preceding minimization technique are the experimental compressive strengths of the $[\pm 15]_s$ angle-ply laminates. These laminates were manufactured for testing. Four sheets of conditioned 3mm x 610mm x 610mm Douglas-fir heartwood veneer were laminated with a phenol-formaldehyde resin (PF 355H) based adhesive and pressed using a 1600 KN capacity Pathex 760 x 760 mm^2 hot-press at 150^0C and 1.38 MPa pressure for 6 minutes. The sheets were oriented in a symmetrical layout with the outside veneer angle at $+15^0$ to the longitudinal direction and the inside two veneers at an angle of -15^0 to the longitudinal direction. Naturally, the grain angle of each sheet was not uniform, so a visual average was used. The 15^0 angle was established using a protractor aligned with the average grain. The lathe checks of the bottom two veneers faced and opposed that of the top two veneers to avoid cupping.

The specimens (nominally 11 x 19 x 40 mm^3) were cut from 4 separate boards. They were prepared in the same manner as described in section 5.3.1.1. (ie. with right angles and a small divot at center cross section to minimize eccentricities). A length of 40 mm was established in accordance with ASTM D198 for short columns with no lateral support.

The specimens were tested in compression using the same test method described in Section 5.3.1.2.

5.4.2.2 Experimental Results

The mechanical properties of the $[\pm 15]_s$ angle-ply laminates are summarized in Table 5.9. Figure 5.20 demonstrates the range of results for the stress-strain behaviour showing 14 of the total 31 curves including the strongest, weakest, stiffest, least stiff and the computed average. The results have been zero-adjusted to remove nonlinearites at the curve origin per section 5.3.1.3. The behaviour is predominantly ductile, although ultimate failure was generally abrupt. Visual inspection of the failed specimens (Figure 5.21) showed a combination of tension perpendicular-to-grain and/or in-plane shear failure.

Table 5.9 - Compression Properties of $[\pm 15]_s$ Angle-ply Laminate

Statistics	Count	Elastic Modulus	Ultimate Stress
Average (MPa)	31	6505.27	51.65
COV (%)	-	26.28	15.39

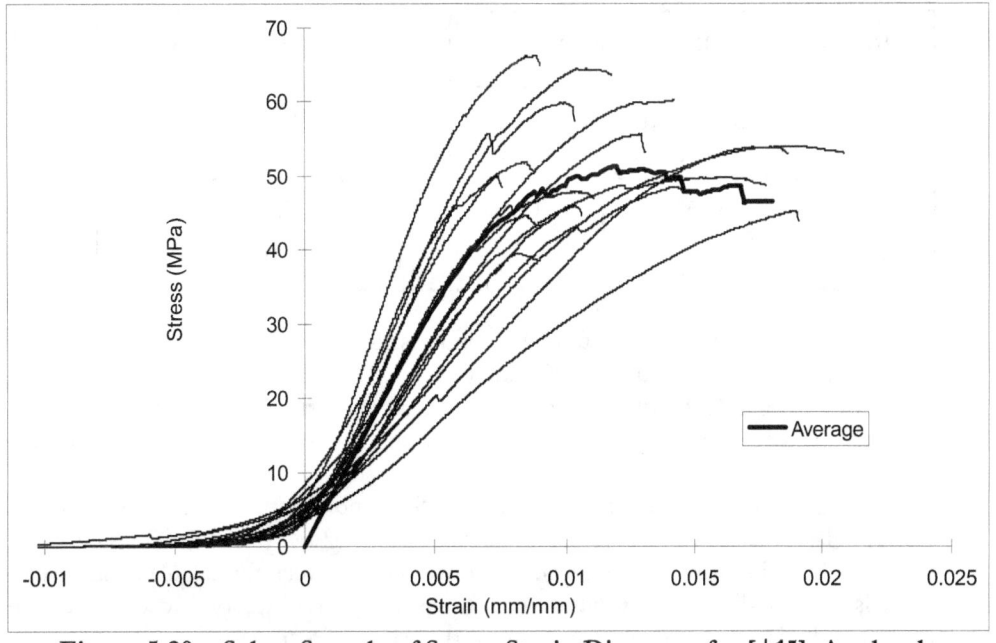

Figure 5.20 - Select Sample of Stress-Strain Diagrams for $[\pm 15]_s$ Angle-ply Laminates

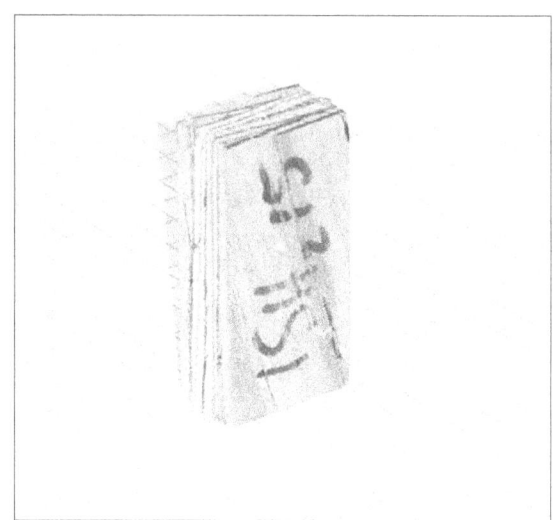

Figure 5.21 - [±15]ₛ Angle-ply Laminate
Compression Specimen

5.4.3 Minimization Results

5.4.3.1 Minimization using the 2 Dimensional Model

The results of the minimization technique using the 2-dimensional model to calculate compression strength of the $[\pm 15]_s$ angle-ply laminate, are summarized in Table 5.10. (Details of the 2 dimensional model were outlined in section 4.2 and a thorough description of the application of the program is deferred to Chapter 6).

Table 5.10 - Results of Nonlinear Minimization using 2-dimensional Model

Statistics	$[\pm 15]_s$ Compression		S	G	F_{12}
	Experimental (MPa)	Simulated (MPa)	(MPa)	(MPa)	(MPa^{-2})
Mean	51.57	51.65	5.99	232.8	5.14×10^{-04}
COV (%)	14.42	15.39	11.7	17.76	71.73

The model predicts the average ultimate compression strength of a $[\pm 15]_s$ angle-ply laminate to be 51.65 MPa. This is very close to the experimental result, 51.57 MPa, as would be expected since the parameters were fitted using this experimental data. The predicted variability is also very accurate, with predicted and experimental coefficients of variations of 15.39 percent and 14.42 percent, respectively.

The average values and variability obtained for S, G and F_{12} appear reasonable. Shear and shear modulus for Douglas-fir clear wood have been reported by Bodig and Jayne (1993) as 6.4 MPa and approximately (using a 14:1 ratio for E_1:G) 800 MPa. The difference between the calculated and published results for the latter is attributed to the thin nature of strands as opposed to solid wood specimens. The interaction parameter F_{12} is more difficult to evaluate as there is no directly comparable data in the literature for wood strands. As such, one can evaluate it based on its conformity to a deterministic evaluation of the stability bounds (Equation 2.8) as follows:

From Tables 5.1 and 5.8, the average strength values are:

X_c=67.32 MPa ; Y_c=15.37 MPa ; X_t = 68.77 MPa ; Y_t =1.91 MPa

The tensile strengths, X_t and Y_t, are first adjusted for size effect. For tension parallel-to-grain, the shape parameter of the 2-parameter Weibull distribution is calculated from maximum likelihood approach as β=4.23 . Strength is adjusted from the length of the test specimen (L_1) to the length of one strand in the prediction specimen (L_2) using Equation 5.6 as follows:

Given:

$$X_{t1} = 68.77 \ \text{MPa}$$

$$L_1 = 50.8 \ \text{mm}$$

$$L_2 = 40.0 \ \text{mm}$$

$$\beta = 4.23$$

then: $\quad X_{t2} = X_{t1}\left(\dfrac{L_1}{L_2}\right)^{\frac{1}{\beta}} = 68.77\left(\dfrac{50.8}{40.0}\right)^{\frac{1}{4.23}} = 72.8 \ \text{MPa}$

For tension perpendicular-to-grain, k=6.66 and strength is adjusted from the volume of the test specimen (V_1) to the volume surrounding one integration point (V_2) using Equation 5.5:

Given:

$$Y_{t1} = 1.91 \ \text{MPa}$$

$$V_1 = 19.09 \times 17.35 \times 152.0 = 50{,}344.2 \ \text{mm}^3$$

$$V_2 = 2.375 \times 5.0 \times 2.55 = 30.28 \ \text{mm}^3$$

$$\beta = 6.66$$

then: $\quad X_{t2} = X_{t1}\left(\dfrac{V_1}{V_2}\right)^{\frac{1}{\beta}} = 1.91\left(\dfrac{50344.2}{30.28}\right)^{\frac{1}{6.66}} = 5.82 \ \text{MPa}$

Using Equation (2.8), the stability bounds are found to be $\pm 1.51 \times 10^{-3}$ MPa^{-2}. The result of 5.14×10^{-4} MPa^{-2} falls within these bounds.

As a visual reference, the cumulative probabilities of the 31 experimental ultimate strengths and 500 simulated strengths using the finalized statistical parameters for S, G and F_{12} have been plotted in Figure 5.22. As expected, the distributions are very close. A log-normal distribution to the experimental data was also plotted for reference. Comparing the three curves, the experimental data tends to deviate slightly at the lower end of the curve, possibly due to manufacturing or test imperfections. Overall, however, the model predictions fit the experimental data extremely well.

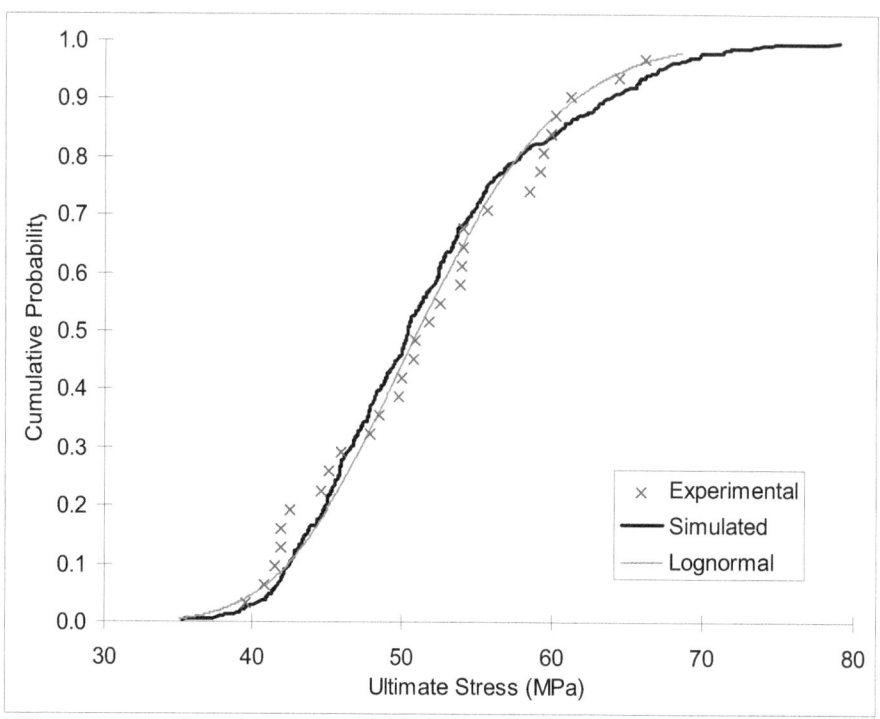

Figure 5.22 - Cumulative Probability Distribution of [±15]ₛ Laminate in Compression (using 2-D Finite Element Model)

500 stress-strain curves for the [±15]ₛ compression laminates were computer generated and compared with the experimental curves. The results are given in Figure 5.23. For clarity, only 5 curves of each (experimental and simulated) have been shown, which represent the average as well as the upper and lower bound for both stiffness and strength. The average curves were obtained by sequentially averaging stress along constant lines of strain. It is noted that the average curves serve only as a reference and is not indicative of material response, per se ; particularly within the higher strain range where the average values are based on fewer points because of varying total strain range. The simulated curves show the loading increments used in the model.

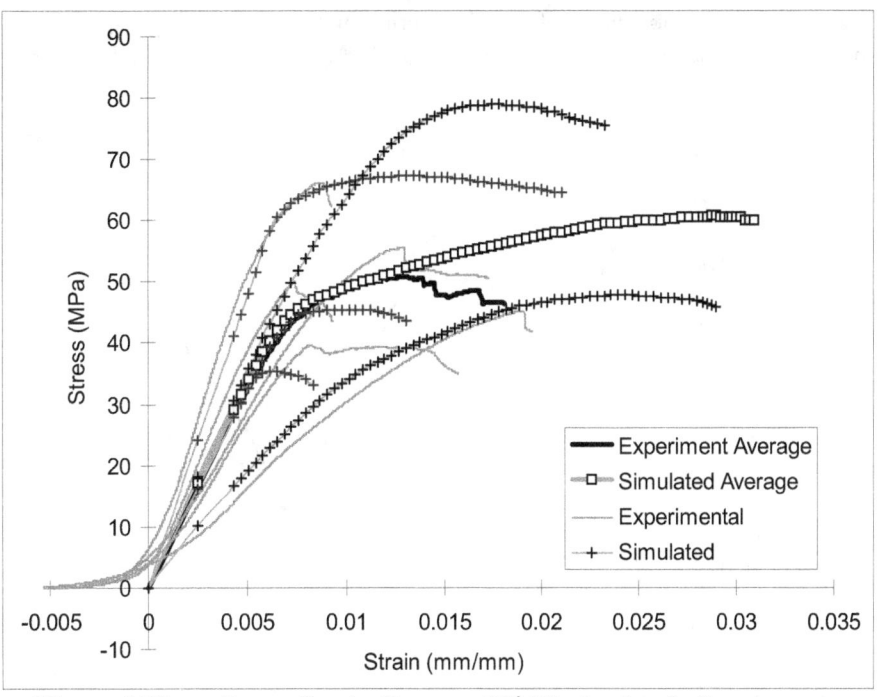

Figure 5.23 - **Stress-Strain Curves of [±15]$_s$ Angle-ply Laminate in Compression (using 2-d Finite Element Model)**

The simulated curves clearly lay within the experimental bounds. Furthermore, they capture the nonlinear behaviour of the [±15]$_s$ laminates.

5.4.3.2 Minimization using 3-Dimensional Model

The results of the minimization technique using the 3-dimensional model (described in section 4.3), are summarized in Table 5.11.

Table 5.11 - Result of Nonlinear Minimization using 3-dimensional F.E. Model

Statistics	[±15]$_s$ Compression		S	G	F$_{12}$
	Experimental (MPa)	Simulated (MPa)	(MPa)	(MPa)	(MPa^{-2})
Average	51.57	51.28	11.4	392.2	1.067x10^{-03}
COV (%)	14.42	13.84	21.8	21.7	8.78

Again, the average values and variability obtained for S, G and F$_{12}$ appear reasonable. The shear strength, in this case however, is twice that found using the 2 dimensional model. It should be understood that, although they compare well with published data, these values serve to calibrate the model and are expected to produce different results for different models. The interaction parameter is also slightly higher, although still within the deterministic bounds.

The cumulative probabilities of the 31 experimental ultimate strengths and 500 simulated strengths using the 3-D model have been plotted in Figure 5.24. As in the 2-D model the experimental distribution is well predicted.

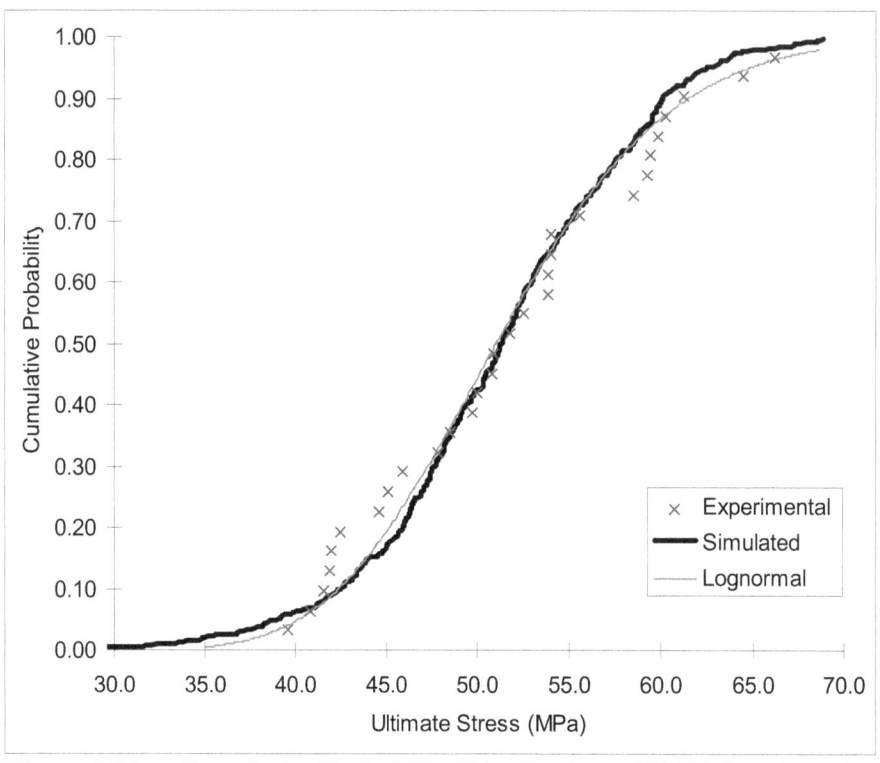

Figure 5.24 - **Cumulative Probability Distribution of [±15]ₛ Laminate in Compression (using 3-D Finite Element Model)**

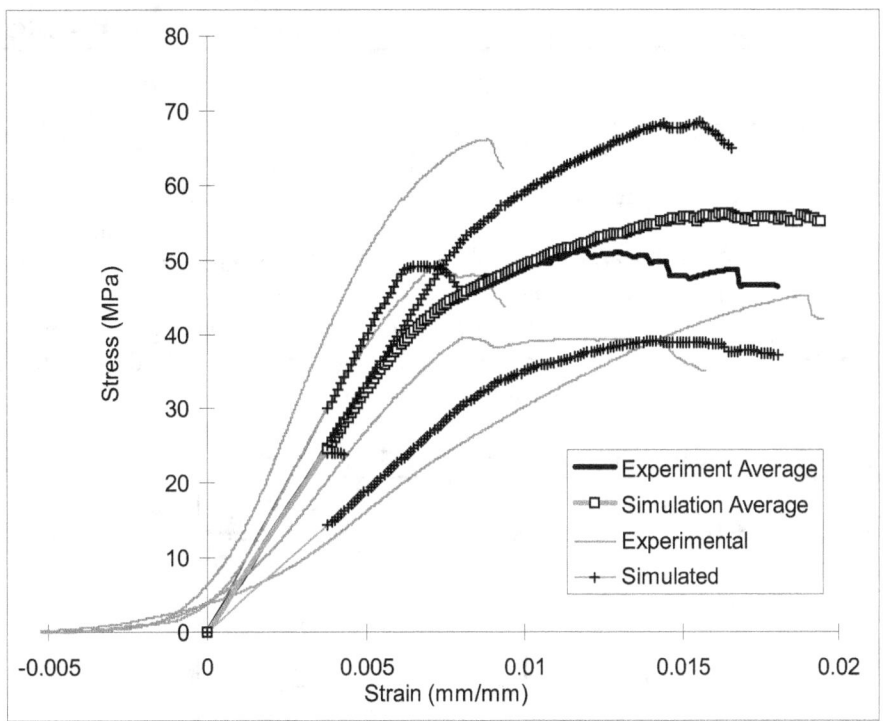

Figure 5.25 - **Stress-Strain Curves of [±15]ₛ Angle-ply Laminate in Compression (using 3-d Finite Element Model)**

Again, 500 stress-strain curves for the [±15]ₛ compression laminates were computer generated - this time using the 3 dimensional model, and compared with the experimental curves. The results are given in Figure 5.25.

It is noted that the programs are adept at identifying mode of failure. Although precise initial mode of failure is difficult to ascertain from experiments due to microscopic cracks, it would appear from the stress - strain curves above that ductile behaviour is prevalent prior to ultimate tensile collapse as is predicted by the simulated results. In both cases, yielding occurs throughout the specimen due to predominant compressive stresses until a combination of tension perpendicular-to-grain and shear stresses cause abrupt brittle failure.

6.1 INTRODUCTION

Having obtained all necessary strength parameters described in the previous chapter, the model is now in a complete form. Both the 2 dimensional and 3 dimensional program can be utilized to predict the nonlinear load - deflection behaviour of structural laminates up to and beyond failure. Within the objectives of this thesis, however, it remains to determine its accuracy. It is the purpose of this chapter to investigate the ability of the present model to reproduce experimental findings for a variety of combinations of geometrical and loading configurations on selected laminates. A total of eighteen categories were investigated as outlined in Table 6.1.

Table 6.1 - Categories of Model Verification Tests

Laminate	Compression	Tension	Bending
$[0/0/0]_s$	-	-	2 - D, 3 - D
$[90/90/90]_s$	-	-	2 - D, 3 - D
$[\pm15]_s$	2 - D, 3 - D	2 - D, 3 - D	2 - D, 3 - D
$[\pm30]_s$	2 - D, 3 - D	2 - D, 3 - D	2 - D, 3 - D
Parallam®	-	3 - D	3 - D

This chapter is broken into three sections according to loading regime investigated. We begin by looking into simple uniaxial analyses; compression and tension, and progress to bending behaviour investigation. A preliminary investigation into modeling Parallam®, PSL in tension and bending is provided thereafter.

6.2 COMPRESSION VERIFICATION

The compressive behaviour of the $[\pm15]_s$ angle-ply laminate was utilised for the minimization process in the foregoing chapter. Although the good results obtained from this process do attest to the accuracy of the program, it would be redundant to discuss the findings here. The compression verification analysis will therefore focus only on the results for the $[\pm30]_s$ angle-ply laminates.

6.2.1 Numerical Analysis

A 2 dimensional and 3 dimensional analysis were performed to simulate the stress - strain response of the $[\pm30]_s$, angle-ply laminates subjected to compressive loading. The geometrical properties, finite element mesh and boundary conditions are given in Figure 6.1. The finite element mesh properties are outlined in Table 6.2.

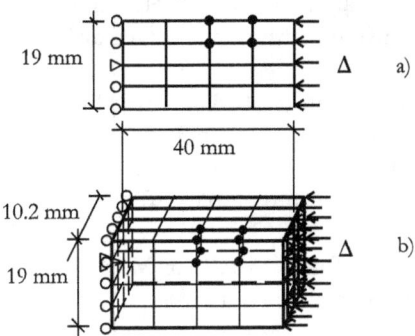

Figure 6.1 - Finite Element Mesh for [±30]ₛ Angle Ply Laminates in Compression: a) 2 dimensions b) 3 dimensions

Table 6.2 - Finite Element Mesh Properties of Angle Ply Laminate Uniaxial Analyses

Finite Element Mesh Properties	2 Dimensional Model	3 Dimensional Model
Number of nodal points	25	125
Number of degrees of freedom	50	375
Number of elements	16	64

This mesh size was chosen for all uniaxial tests following a preliminary parametric study which is discussed in Appendix B.

The 2 dimensional model utilized 4-node elements. Referencing Figure 6.2, each element was comprised of 4 layers where each layer represented an individual ply with a given ply angle. Using the assumptions of the classical lamination theory, outlined in Section 4.2.2.3, no slip was assumed between plies and strains are established in terms of the mid-surface strain. Therefore, using a 2x2 Gauss quadrature rule, there were 4 Gaussian points per element to establish the strains and 16 different stress points (4 points x 4 layers) per element or a total of 256 points per specimen monitored by the yield/failure criterion throughout the entire stress path.

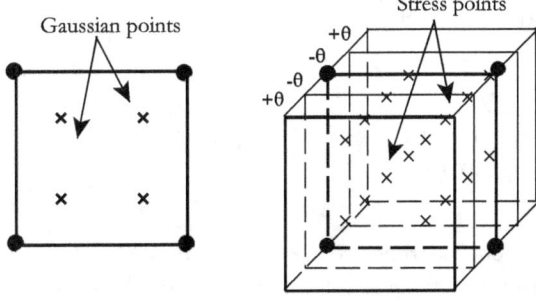

Figure 6.2 - Explanatory Depiction of 2 Dimensional Element

The 3 dimensional model employed 8-node brick elements as outlined in Section 4.3. Using a 2x2x2 quadrature rule, there were 8 integration points per element or a total of 512 integration points per specimen. As a result, the 3 dimensional program involved approximately 8 times more computer time.

As the post-failure behaviour was sought, the laminate was loaded by prescribing an increasing value of displacement (as opposed to load). For each increment in displacement, the nonlinearities in the equilibrium equations were resolved using the modified Newton-Raphson iterative procedure. This displacement approach avoids problems with global stiffness matrix singularities beyond peak load.

To assist in the making of input files for COMAP, an MS Excel® Visual Basic® macro was created. A property input dialog was used to manage the rather large number of statistical parameters required for COMAP. To demonstrate the use of this for the present analysis, the material stiffness and strength parameters for input into the 2 dimensional program are summarized in 'Windows' format in Figure 6.3.

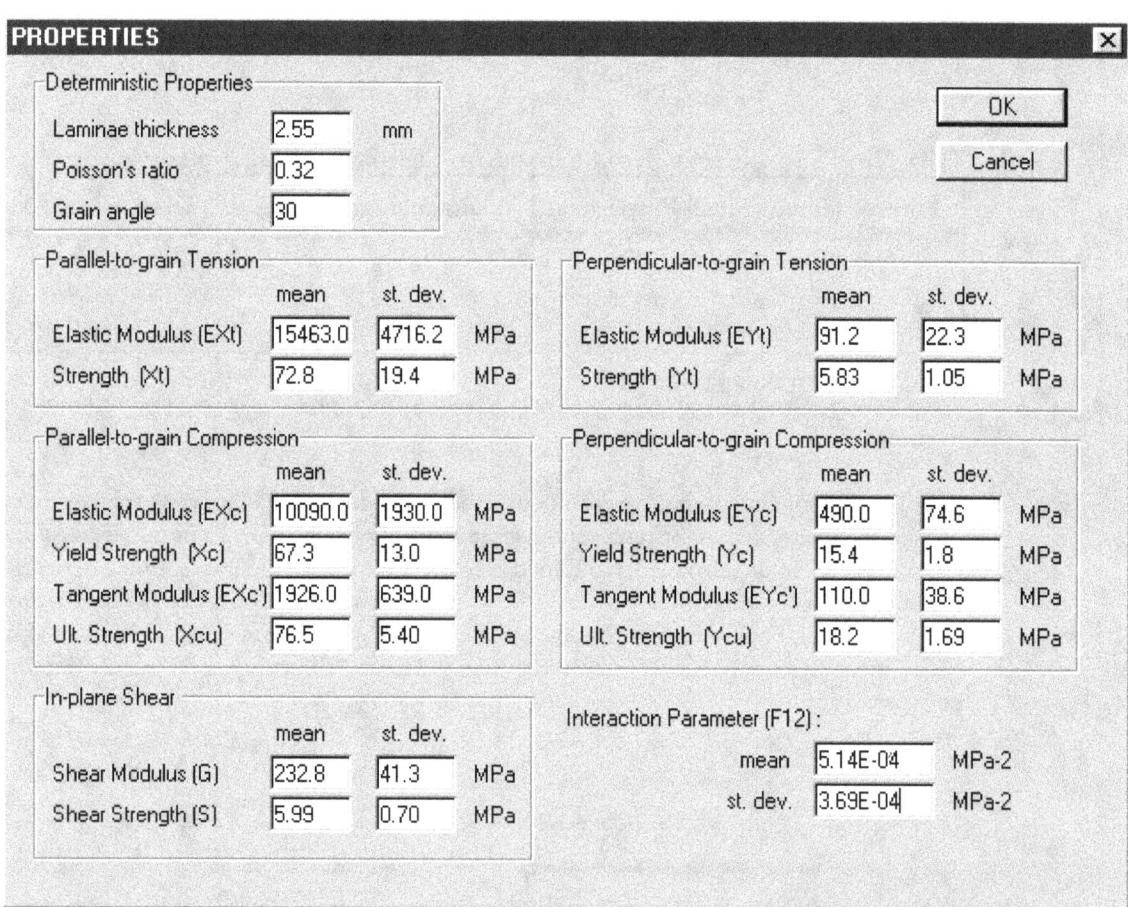

Figure 6.3 -   Input Parameters for [±30]ₛ Angle Ply Laminate in Compression (2 Dimensional Program)

It is noted that the tensile strengths given in Figure 6.3 have been adjusted from those given in Table 5.6. The reason for this is that prior to implementing the values into the finite element code, adjustments must first be made for size effect as described in Section 5.3.3. The perpendicular-to-grain tension strength was adjusted from representing that of the tested volume to that of the tributary area for one Gauss point. Longitudinal strength was adjusted however, from the experimental gauge length to the model gauge length of 40 mm. This is exactly the same calculation using the same values as was outlined in Section 5.4.3.1. for the $[\pm 15]_s$ laminate in compression and will not be repeated here.

The calibrated factors, S, G and F_{12}, as well as the tensile strength Y_t differ for the 3 dimensional program from Figure 6.3. These input parameters are presented in Table 6.3. The mean value of Y_t is different as a result of size effect (ie. due to the smaller volume surrounding each Gaussian point for the 3 dimensional model). The coefficient of variation of Y_t is assumed to be the same for both models.

Table 6.3 - Input Properties of 3 D. Program Differing from those of 2 D. Program

Property	Mean	Standard Deviation
Shear Strength, S (MPa)	11.4	2.6
Shear Modulus, G_{12} (MPa)	390.0	89.7
Interaction Parameter, F_{12} (MPa^{-2})	1.067×10^{-03}	9.368×10^{-05}
Perpendicular-to-grain Tension Strength, Y_t (MPa)	6.46	1.14

The elastic moduli associated with the through-thickness direction, E_3, G_{13} and G_{23}, have been estimated through general moduli relationships cited in Bodig and Jayne (1993) as

$$E_3 : E_2 \approx 1.6 : 1$$
$$G_{13} : G_{12} : G_{23} \approx 10 : 9.4 : 1$$

The major Poisson's ratios in the out-of-plane dimension are estimated by approximate relationships for Douglas-fir from Bodig and Jayne (1993):

$$\nu_{13}/E_1 = 2.6 \times 10^{-05} \text{ MPa}$$
$$\nu_{23}/E_2 = 4.9 \times 10^{-04} \text{ Mpa}$$

and the reciprocal relationships of Equation 3.10 are invoked for the reciprocal identities. These elastic moduli are 100 percent correlated with the generated random value; for example, for each random value chosen for E_2, a new value for E_3 is calculated based solely on the given ratio.

The program is formulated for stochastic analyses. All strength and stiffness properties were assumed to vary between layers, with the exception of tension perpendicular to grain strength which was instead regenerated for each integration point. This strength was treated differently because in addition to yielding good results, material behaviour in this configuration is believed to be more in keeping with ideal brittle fracture theory. In this configuration, the within member variation is less likely to be correlated as it is, for example, for tension parallel to grain (Lam and Varoglu 1990). The failure mechanism is more likely 'perfectly brittle', failing completely when fracture occurs at the weakest point. The material flaws are assumed to be randomly distributed.

6.2.2 Experimental Tests

For comparison to the analytical data, $[\pm 30]_s$ angle ply specimens (nominally 19 x 10.5 x 40 mm^3) were fabricated, prepared and tested under uniaxial compression in exactly the same manner as the $[\pm 15]_s$ laminates described in Section 5.4.2.

6.2.3 Results

The results of both the numerical simulations and experiments are summarized in Table 6.4 and Figures 6.4 and 6.5.

Table 6.4 - Experimental and Simulated Data for [±30]$_s$ Angle Ply Laminates in Compression

Statistic	Experiment	Simulation (Count = 500)			
	(Count = 39)	2 d	(% error)	3 d	(% error)
Elastic Modulus Mean (MPa)	3052	2273	(25.5)	2742	(10.2)
Elastic Modulus COV (%)	27.5	13.3	(51.6)	8.3	(69.8)
Strength Mean (MPa)	20.3	19.5	(4.0)	21.1	(3.8)
Strength COV (%)	11.1	9.6	(13.9)	13.8	(23.9)

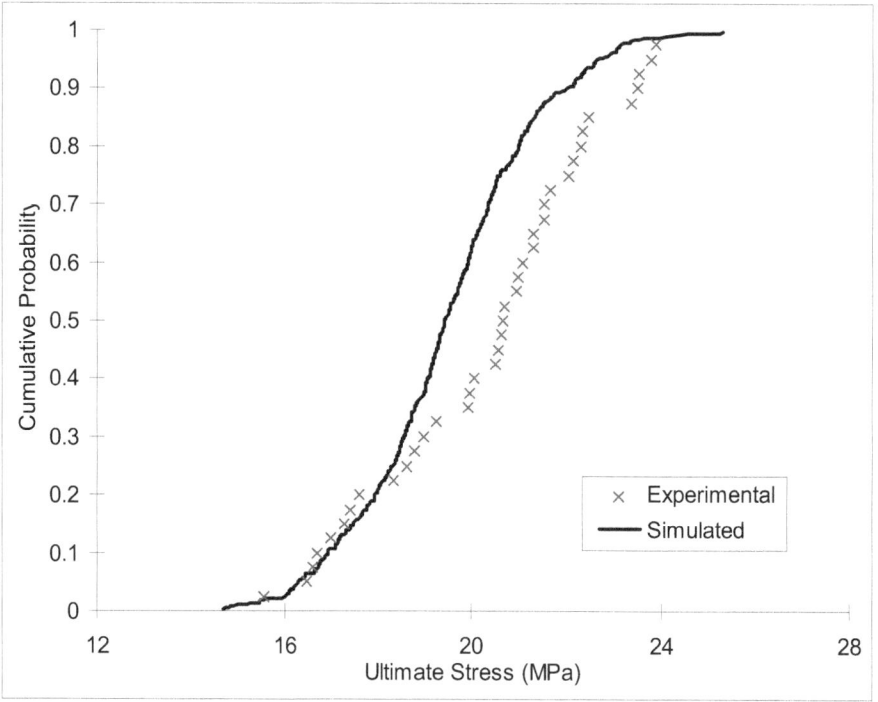

Figure 6.4 - Cumulative Probability Distribution of [±30]$_s$ Laminate in Compression (2 - Dimensional Model)

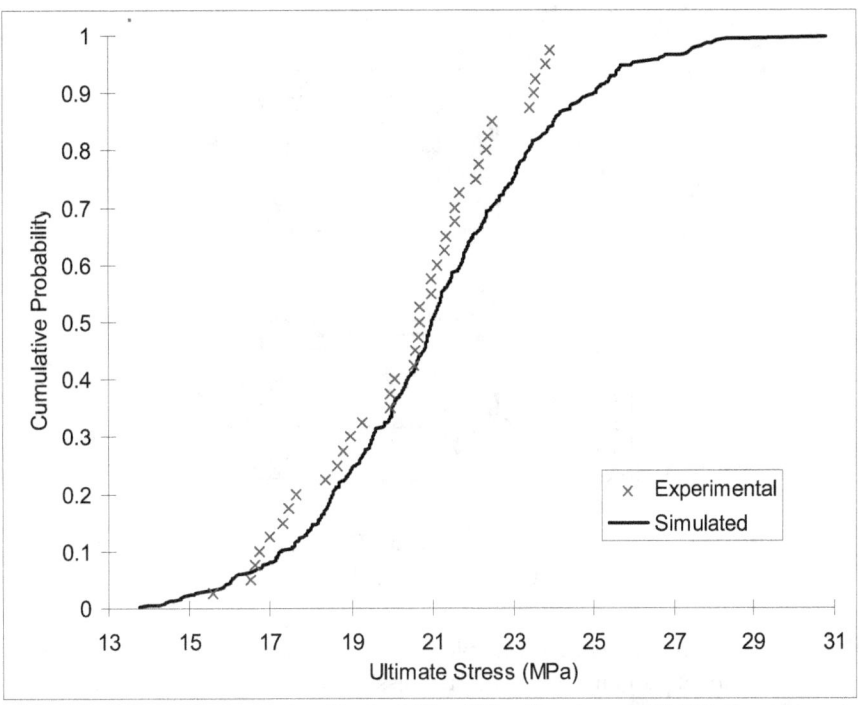

Figure 6.5 - **Cumulative Probability Distribution of [±30]$_s$ Laminate in Compression (3 - Dimensional Model)**

As evident from Table 6.4, the average value of the ultimate strengths is well predicted (maximum percent error of 4.0 percent). The average initial stiffness is underestimated for the 2 dimensional program yet quite accurate for the 3 dimensional program. The difference in the two programs is due to the lower shear stiffness calibrated for the 2 dimensional than for the 3 dimensional program (232.8 MPa vs. 390.0 MPa, respectively).

Referencing Figures 6.4 and 6.5, the upper region of the cumulative probability distribution curve is slightly under-predicted by the 2 dimensional model, yet over-predicted by the 3 dimensional model. This is partly a result of the number of stress points for each model. The 3 dimensional model, having many more points for stress redistribution after localized failure, seemed to be better able to capture nonlinear behaviour and as a result produced higher laminate capacities. This may also account for the higher variability of the 3 dimensional model.

Stress Redistribution
To illustrate how both programs manage stress redistribution upon integration point failure, as discussed in Section 4.4.3., the parallel-to-grain stress (σ_1) at prescribed points within a specimen (shown in Figure 6.6) have been tracked up to and beyond peak load. The stress-strain diagrams for each point, together with the overall longitudinal laminate behaviour, is given in Figures 6.7 and 6.8 for the 2 dimensional and 3 dimensional programs, respectively. Overall laminate stress is computed as the total applied force divided by the cross-sectional area of the specimen.

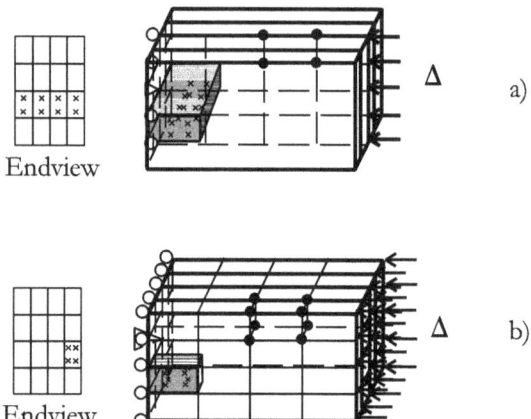

Figure 6.6 - **Location of Sample Points used to Show Stress Redistribution: a) 2 Dimensions b) 3 Dimensions**

The location of the points which are plotted are marked as small crosses in cross section in Figure 6.6. For the 2 dimensional program, a total of 16 stress points are tracked. These are the 4 stress points per 4 layers of one element. For the 3 dimensional program, a total of 8 points are plotted which are simply the 8 Gaussian points of one element (one layer thick only).

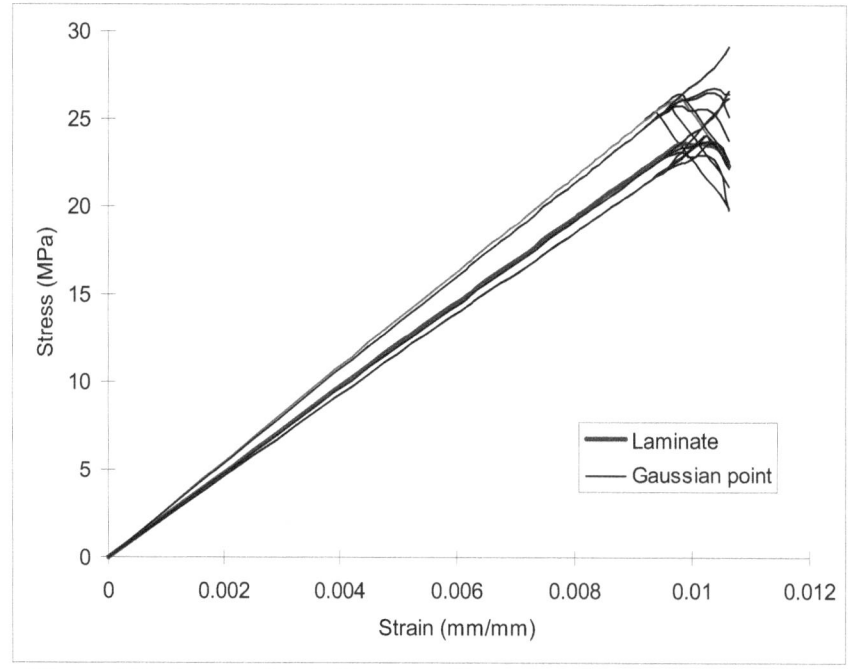

Figure 6.7 - Stress - Strain Diagram Depicting Stress Redistribution Upon Stress Point Failure (2 Dimensions)

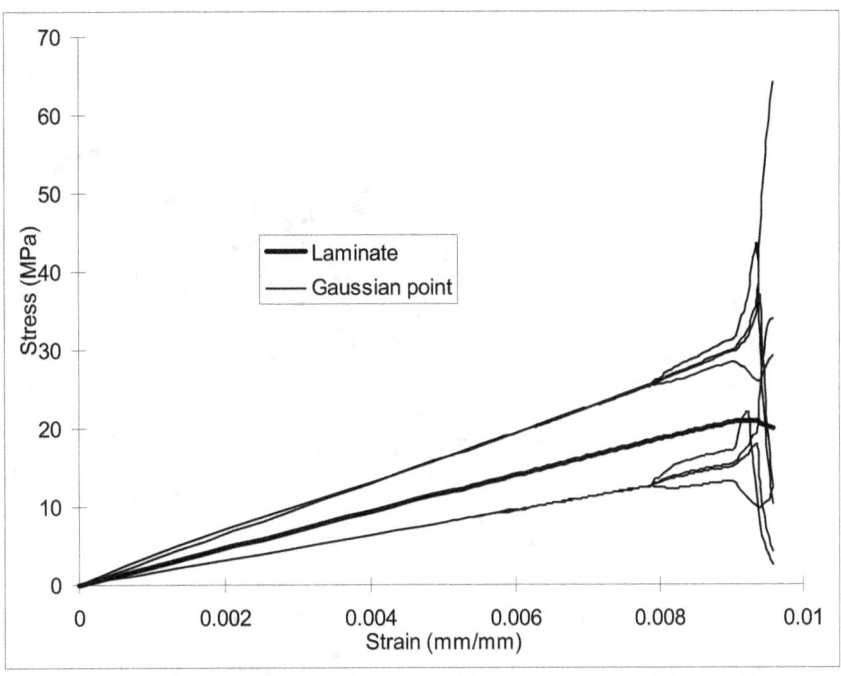

Figure 6.8 - Stress - Strain Diagram Depicting Stress Redistribution Upon Stress Point Failure (3 Dimensions)

Figures 6.7 and 6.8 show points which have failed in a brittle manner (gradual declining curves) and subsequent reaction of some surrounding points to compensate for this loss of capacity. For each increment in displacement shown, the yield criterion and equilibrium requirements are satisfied through the iterative procedure outlined in Section 4.4.2.

Figure 6.9 and 6.10 attest to both program's capabilities to simulate the full range of stress - strain curves. Following the same format as for Section 5.4.3.1., the curves shown demonstrate the strongest, weakest, stiffest, least stiff, and calculated curve average for both experimental and simulated data.

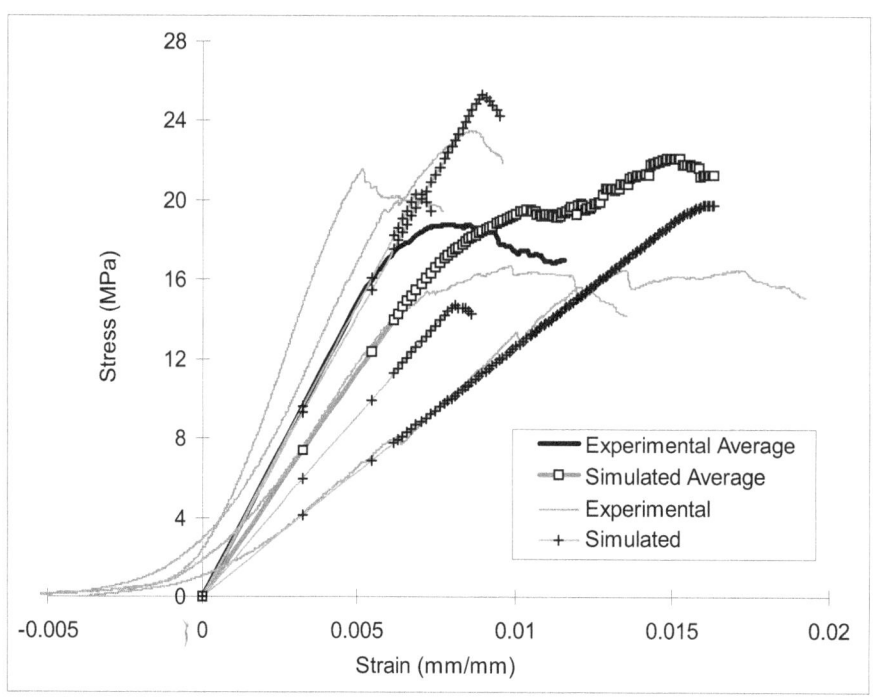

Figure 6.9 -   Stress-Strain Curves of [±30]ₛ Angle-ply Laminate in
Compression (using 2 - Dimensional Model)

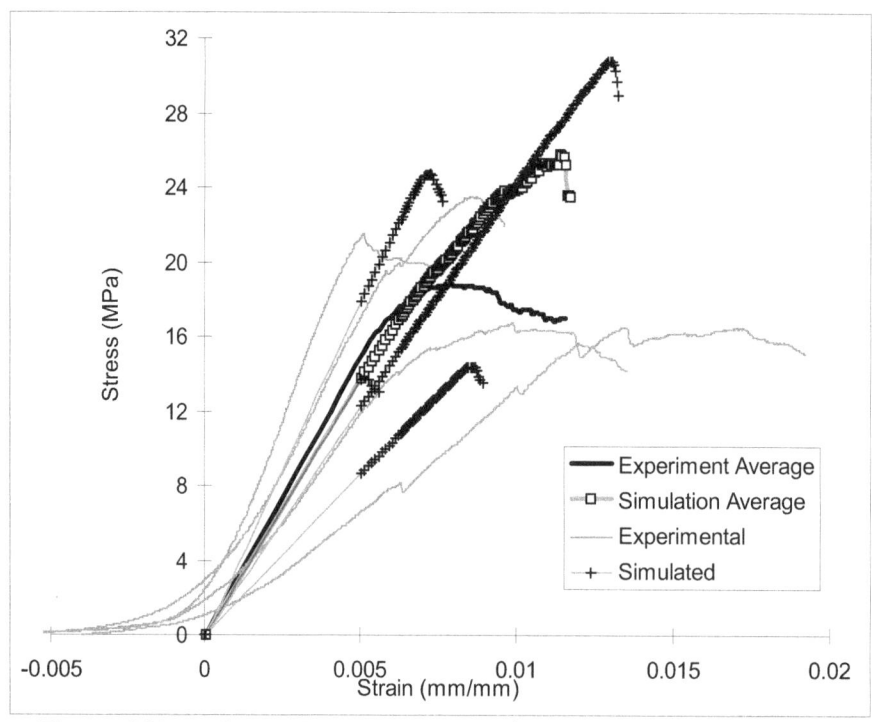

Figure 6.10 -   Stress-Strain Curves of [±30]ₛ Angle-ply Laminate in
Compression (using 3 - Dimensional Model)

6.3   TENSION VERIFICATION

The general performance of the models to predict laminate behaviour under uniaxial tensile loading is investigated in this section using 2 different geometrical configurations: [±15]$_s$ and [±30]$_s$ angle ply laminates.

6.3.1   Numerical Analyses

The 2 dimensional and 3 dimensional computation for constitutive tensile behaviour of the [±15]$_s$ and [±30]$_s$ angle ply laminates follows very closely to that conducted for compressive behaviour discussed in the previous section. The geometrical properties, finite element mesh, and boundary conditions are the same as described in Figure 6.1 save direction of applied displacement and specimen length. The [±15]$_s$ laminates were 120 mm in length and the [±30]$_s$ laminates were 60 mm in length. A longer length was used for the [±15]$_s$ laminates to ensure wood fibres were not continuous from one test grip to the other potentially increasing observed experimental strength.

Given these specimen lengths, the tensile strengths were adjusted for size effect, per section 5.3.3, prior to being used in the programs. The calculation is as follows:

For tension parallel-to-grain, the shape parameter of the 2 - parameter Weibull distribution of the experimental data was calculated from a maximum likelihood approach to be β=4.23 . Strength was adjusted from the length of the test specimen (L$_1$) to the length of one strand in the prediction specimen (L$_2$) using Equation 5.6:

Given:

β = 4.23
X$_{t1}$ = 68.77 MPa
L$_1$ = 50.8 mm
L$_2$ ([±15]$_s$ laminate) = 120 mm ; L$_2$ ([±30]$_s$ laminate) = 60 mm

then for the [±15]$_s$ laminate

$$X_{t2} = X_{t1}\left(\frac{L_1}{L_2}\right)^{\frac{1}{\beta}} = 68.77\left(\frac{50.8}{120.0}\right)^{\frac{1}{4.23}} = 56.1 \ MPa$$

and for the [±30]$_s$ laminate

$$X_{t2} = X_{t1}\left(\frac{L_1}{L_2}\right)^{\frac{1}{\beta}} = 68.77\left(\frac{50.8}{60.0}\right)^{\frac{1}{4.23}} = 66.1 \ MPa$$

The result is valid for both 2 and 3 dimensional analyses. For tension perpendicular-to-grain, β=6.66 and strength is adjusted from the volume of the test specimen (V$_1$) to the volume surrounding one integration point (V$_2$) (which is different for both programs) using Equation 5.5:

Given:

β = 6.66
Y$_{t1}$ = 1.91 MPa
V$_1$ = 19.09 x 17.35 x 152.0 = 50344.2 mm^3

For the 2 dimensional program:

$$V_2 \ ([\pm15]_s \text{ laminate}) = 2.375 \times 15.0 \times 2.55 = 90.6 \text{ mm}^3 \ ; \ Y_{t2} = \left(\frac{50344.2}{90.8}\right)^{\frac{1}{6.66}} = 4.9 \text{ MPa}$$

$$V_2 \ ([\pm30]_s \text{ laminate}) = 2.375 \times 7.5 \times 2.55 = 45.4 \text{ mm}^3 \ ; \ Y_{t2} = \left(\frac{50344.2}{45.4}\right)^{\frac{1}{6.66}} = 5.5 \text{ MPa}$$

and for the 3 dimensional program:

$$V_2 \ ([\pm15]_s \text{ laminate}) = 2.375 \times 15.0 \times 1.27 = 45.4 \text{ mm}^3 \ ; \ Y_{t2} = \left(\frac{50344.2}{45.4}\right)^{\frac{1}{6.66}} = 5.5 \text{ MPa}$$

$$V_2 \ ([\pm30]_s \text{ laminate}) = 2.375 \times 7.5 \times 1.27 = 22.7 \text{ mm}^3 \ ; \ Y_{t2} = \left(\frac{50344.2}{22.7}\right)^{\frac{1}{6.66}} = 6.1 \text{ MPa}$$

Coefficient of variation (COV) is assumed to remain the same as found in the experiment. Hence, the input parameters for the 2 dimensional and 3 dimensional programs are as outlined in Table 6.5.

The elastic moduli associated with the through-thickness direction, E_3, G_{13} and G_{23}, have been estimated through relationships cited in Bodig and Jayne (1993) as explained in Section 6.2.1.

6.3.2 Experimental Tests

Specimens were cut from boards that were made as described in Section 5.4.2.1. The tension specimens were nominally 19 x 11 x120 mm³ ([±15]$_s$ laminates) and 19 x 10.5 x 60 mm³ ([±30]$_s$ laminates) with minimal dimensional variation (1.4 % maximum in the through thickness dimension).

The specimens were tested using a 250 kN MTS universal test machine equipped with MTS mechanical wedge action grips as described for the procurement of strand tensile data in Section 5.3.2.2. The loading rate was set at 1.8 mm/min. to produce failure in 6 minutes, on average. Tensile failure for both [±15]$_s$ and [±30]$_s$ were typical of that shown in Figure 6.12, where perpendicular-to-grain tension and in-plane shear stresses governed failure. Glue failures were not observed.

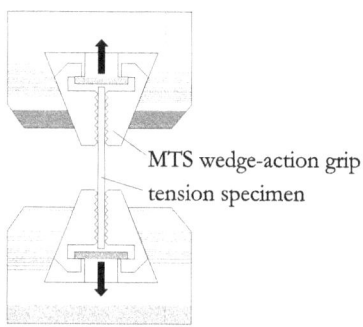

Figure 6.11 - Tension Test Set-up

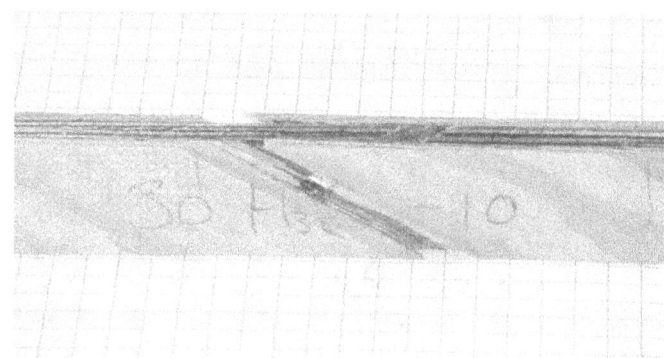

Figure 6.12 -   Tensile Failure of [±30]$_s$ Angle Ply Laminate Specimen

Table 6.5 - Input Properties for $[\pm15]_s$ and $[\pm30]_s$ Angle Ply Laminates in Tension

Property		2 Dimensional		3 Dimensional	
		Mean	St. Dev.	Mean	St. Dev.
Parallel-to-grain Tension	Elastic Modulus (E_{Xt}) (MPa)	15463	4716.2	15463	4716.2
	Strength (X_t) $[\pm15]_s$ (MPa)	56.1	14.9	56.1	14.9
	Strength (X_t) $[\pm30]_s$ (MPa)	66.1	17.6	66.1	17.6
Perpendicular-to-grain Tension	Elastic Modulus (E_{Yt}) (MPa)	91.2	22.3	91.2	22.3
	Strength (Y_t) $[\pm15]_s$ (MPa)	4.9	0.87	5.5	0.96
	Strength (Y_t) $[\pm30]_s$ (MPa)	5.5	0.96	6.1	1.07
Parallel-to-grain Compression	Elastic Modulus (E_{Xc}) (MPa)	10090	1930	10090	1930
	Yield Strength (X_c) (MPa)	67.3	13	67.3	13
	Tang Modulus (E_{Xc}') (MPa)	1926	639	1926	639
	Ultimate Strength (X_c^u) (MPa)	76.5	5.4	76.5	5.4
Perpendicular-to-grain Compression	Elastic Modulus (E_{Yc}) (MPa)	490	74.6	490	74.6
	Yield Strength (Y_c) (MPa)	15.4	1.8	15.4	1.8
	Tangent Modulus (E_{Yc}') (MPa)	110	38.6	110	38.6
	Ultimate Strength (Y_c^u) (MPa)	18.2	1.7	18.2	1.7
Interaction Parameter	F_{12} (MPa^{-2})	5.1×10^{-04}	3.7×10^{-04}	1.1×10^{-03}	9.4×10^{-05}
In plane Shear	Elastic Modulus (G) (MPa)	232.8	41.3	392.2	85.1
	Strength (S) (MPa)	5.99	0.7	11.4	2.6
Poisson's Ratio	ν_{12}	0.32	-	0.32	-

6.3.3 Results

The results of the experimental tests and simulations are summarized in Table 6.6. The tension grips, depicted in Figure 6.11, were of a wedge type such that as the load increased, the measured displacement of the specimen included inherent displacement within the grips. As a result, the load-displacement curve exhibits, incorrectly, a lower stiffness. Unfortunately, laminate stiffness was not measured using an extensometer and so cannot be compared to the simulated data. Regardless, descriptive statistics are included for the simulated elastic stiffness in Table 6.6 for completeness. The 2 dimensional model is less stiff than the 3 dimensional model due to the lower shear stiffness value used as input.

Table 6.6 - Experimental and Simulated Data for $[\pm 15]_s$ and $[\pm 30]_s$ Angle Ply Laminates in Tension

Config.	Statistic	Experiment	Simulation (Count =500)			
		{Count}	2 d	(% error)	3 d	(% error)
$[\pm 15]_s$	Elastic Modulus Mean (MPa)	-	6724.1	-	9128.6	-
	Elastic Modulus COV (%)	-	11.2	-	12.8	-
	Strength Mean (MPa)	39.4 {39}	39.1	(0.6)	36.4	(7.6)
	Strength COV (%)	16.7	20.7	(23.9)	17.5	(4.8)
$[\pm 30]_s$	Elastic Modulus Mean (MPa)	-	2273.0	-	3132.7	-
	Elastic Modulus COV (%)	-	13.3	-	10.85	-
	Strength Mean (MPa)	21.7 {41}	21.3	(1.8)	21.4	(1.4)
	Strength COV (%)	9.0	12.1	(34.4)	13.2	(46.7)

Cumulative probability distributions for the $[\pm 15]_s$ and $[\pm 30]_s$ angle ply laminates in tension are plotted in Figures 6.13 through 6.16. Ultimate stress is predicted relatively well by both models for both configurations (maximum percent error of 7.6%). The larger prediction error of the $[\pm 15]_s$ laminate strength for the 3 dimensional case is attributed to the fact that the 3 dimensional model has more Gaussian points than the 2 dimensional model for potential failure. We consider the stress state of a Gaussian point at incipient failure for the $[\pm 15]_s$ laminate. For purposes of explanation, a likely stress state would be $\{\sigma_1, \sigma_2, \sigma_4\}^T = \{39, -0.5, 4.1\}^T$. From this, it is seen that tension parallel-to-grain governs. If the point is weak in tension perpendicular-to-grain and the layer is weak in tension parallel-to-grain, then the point will fail immediately in a brittle manner. Being a brittle failure, as soon as one point fails, surrounding points (particularly in the weak layer) are heavily stressed and also fail, leading to catastrophic failure. Since the 3 dimensional model has more points, there is a higher probability of having a point which is weak in tension perpendicular-to-grain, and hence, of having overall lower laminate strengths. It is noted that the size adjustment to Y_t for the 3 dimensional case does negate this somewhat; however, one could speculate that the Weibull approach may not be wholly appropriate for multi-axial stress states. A future study could investigate this further.

The $[\pm 30]_s$ laminate, on the other hand, fails numerically in a slightly more ductile manner (see Figure 6.17). A typical stress state of a Gaussian point at failure for the $[\pm 30]_s$ laminate is $\{\sigma_1, \sigma_2, \sigma_4\}^T = \{22, -1.1, 6.2\}^T$. In this case, the conditions of Table 4.1 may not be immediately met and the point may initially yield. Referencing Figure 3.1, when a point yields, it maintains strength and does not put the same burden on surrounding points as would a brittle failure. Thus, a weak point does not have the same impact on a $[\pm30]_s$ laminate as for the $[\pm15]_s$ laminate. For this reason, the 3 dimensional and 2 dimensional models are closer in prediction for the $[\pm30]_s$ laminate.

The coefficient of variation of the $[\pm30]_s$ laminate strength is somewhat high for both models. It is speculated that this occurs as a result of the variability for the fitted parameters, S (in-plane shear) and to a lesser degree, F_{12} (interaction parameter). These parameters were established using the $[\pm15]_s$ laminate in compression (for reasons noted in section 5.4.1), which has a relatively low in-plane shear stress. Considering the stress state at failure given above for the $[\pm30]_s$ laminate, the strong presence of shear stress influences the ultimate laminate strength. If the shear variability is estimated as high, the laminate variability will be high.

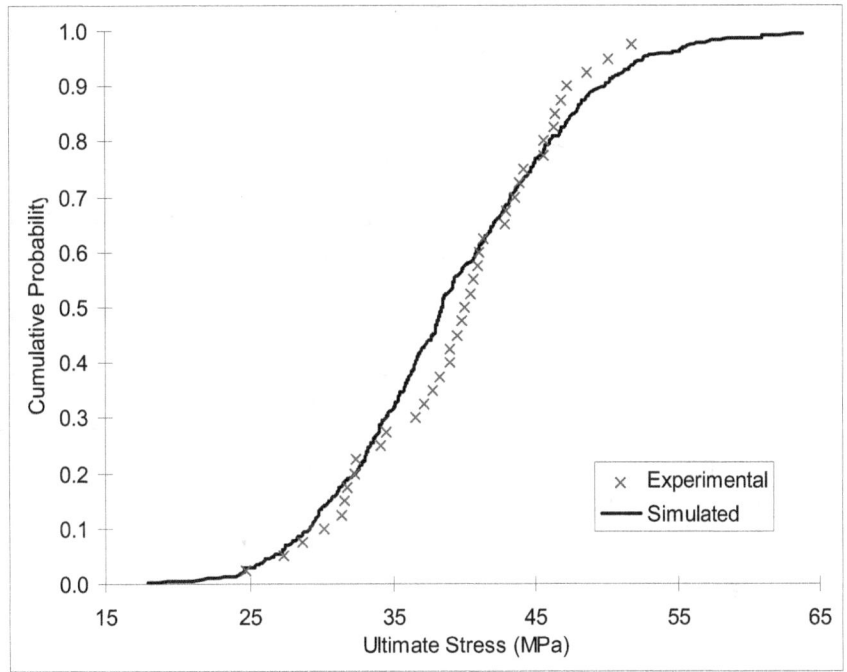

Figure 6.13 -   Cumulative Probability Distribution of $[\pm15]_s$ Laminate in Tension (2 - Dimensional Model)

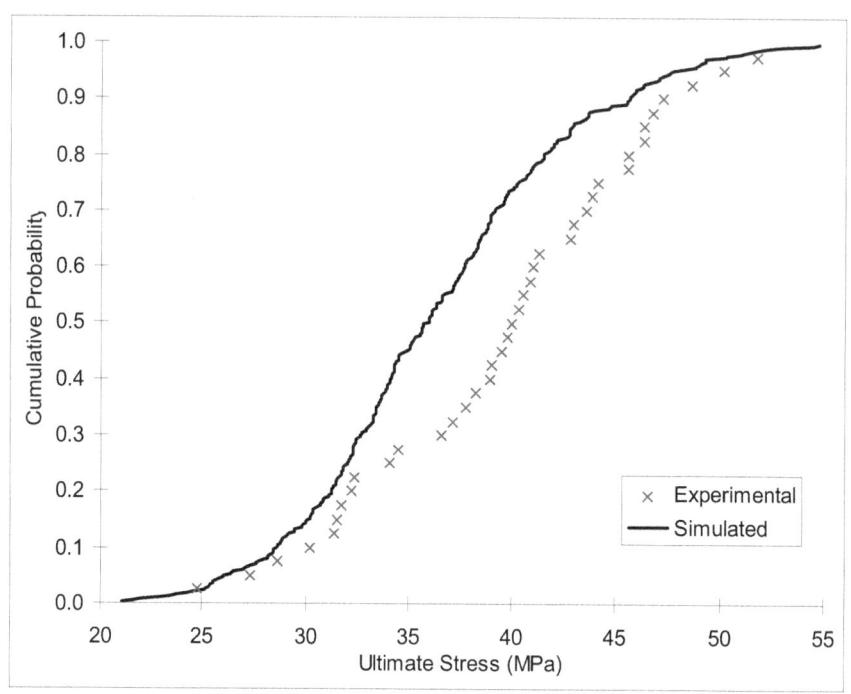

Figure 6.14 -   Cumulative Probability Distribution of [±15]$_s$
Laminate in Tension (3 - Dimensional Model)

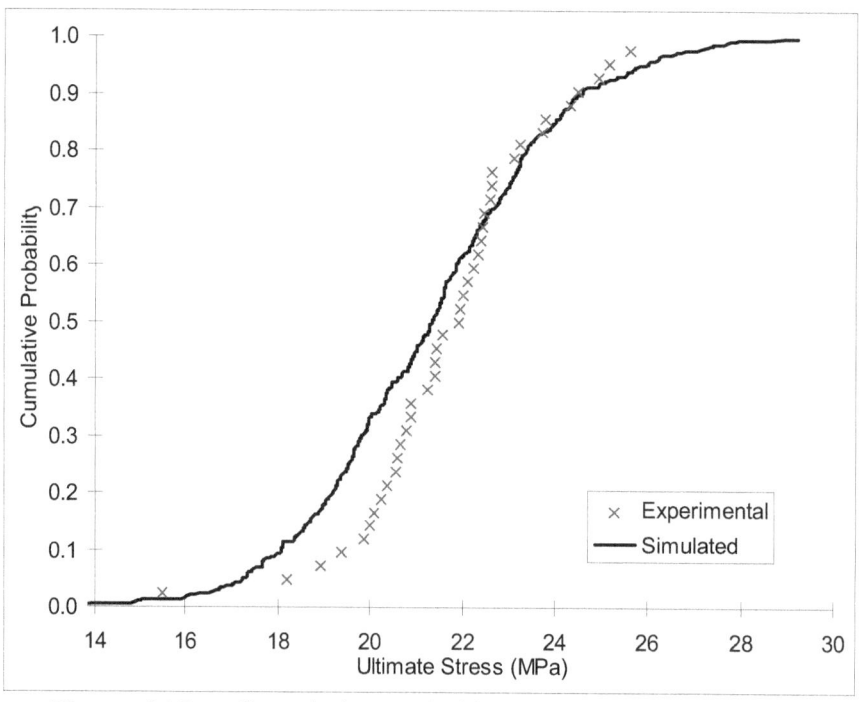

Figure 6.15 -   Cumulative Probability Distribution of [±30]$_s$
Laminate in Tension (2 - Dimensional Model)

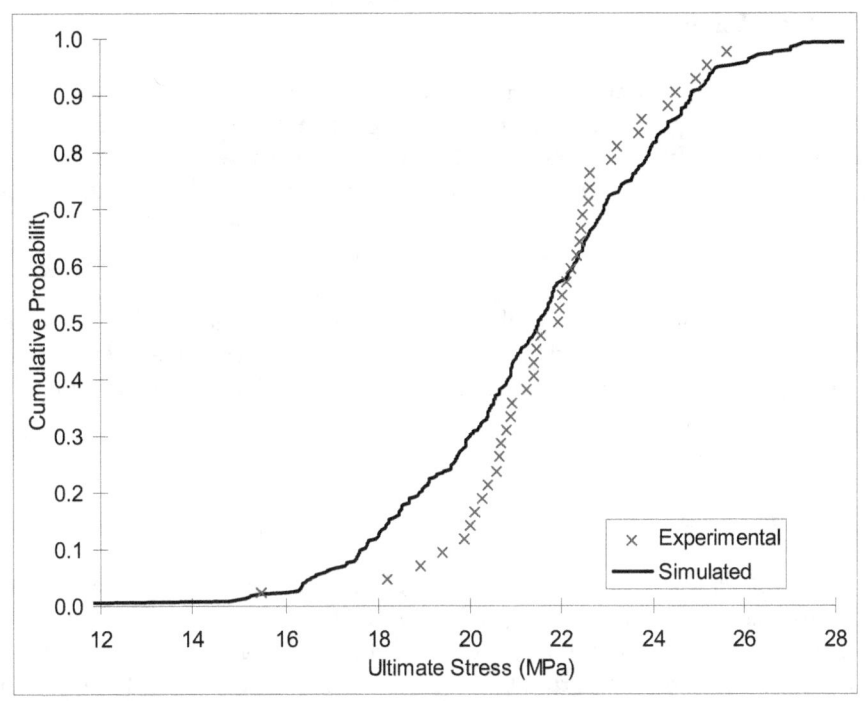

Figure 6.16 - Cumulative Probability Distribution of $[\pm 30]_s$ Laminate in Tension (3 - Dimensional Model)

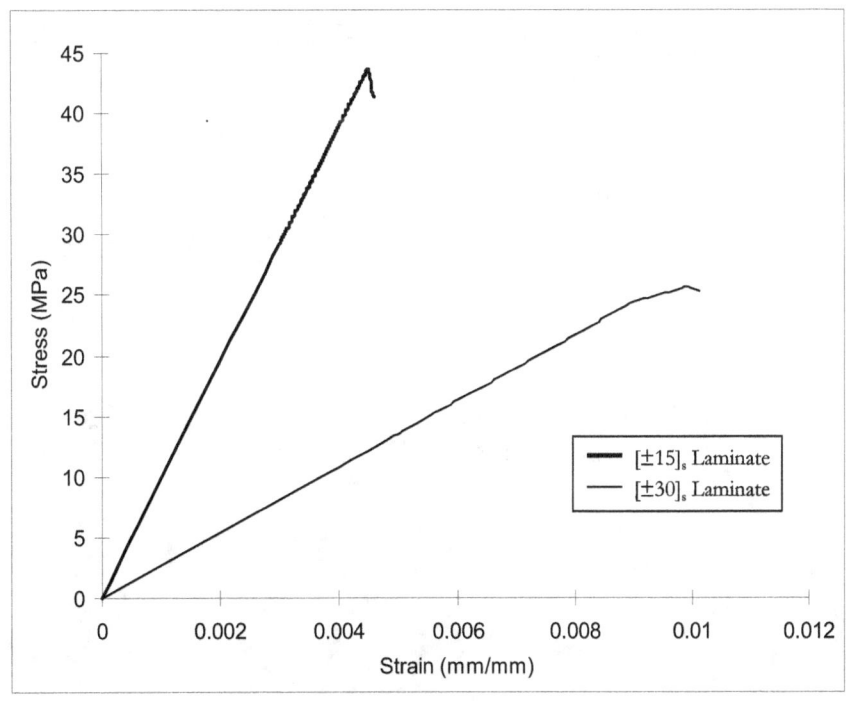

Figure 6.17 - Stress-Strain Curve Comparison of $[\pm 15]_s$ and $[\pm 30]_s$ Laminates in Tension

6.4 BENDING VERIFICATION

In this section, we check the intended purpose of the model - to simulate the mechanical response of a wood strand composite in bending. In bending, all quadrants of the stress field are present simultaneously. Initial specimen failure may be a result of compressive stresses (present in the upper fiber of a positive bending specimen), while ultimate failure may result from predominantly tensile stresses (present in the lower fibers of said specimen). Hence, the bending analysis provides a rigorous check for the simulation model.

This section reports the numerical and experimental methodology, as well as the comparative results for the response of $[\pm15]_s$ and $[\pm30]_s$ angle ply laminates when subjected to 3 point bending. In addition, bending behaviour for laminated veneer specimens (with no grain angle variation - 0 and 90 degrees) is explored and discussed.

6.4.1 Numerical Analyses

Figure 6.18 illustrates the geometric layout and corresponding finite element mesh of the 2 and 3 dimensional models for the $[\pm15]_s$ and $[\pm30]_s$ angle ply laminates under 3 point bending. The same mesh was used for the 0° and 90° laminates with the exception that these laminates were 6 plys in through thickness. The mesh properties for both programs are outlined in Table 6.7. As shown, the size of the 2 dimensional program is relatively unaffected by the number of layers in the laminate. However, the 3 dimensional program becomes unmanageably large very quickly due to the number of elements, even with small specimens.

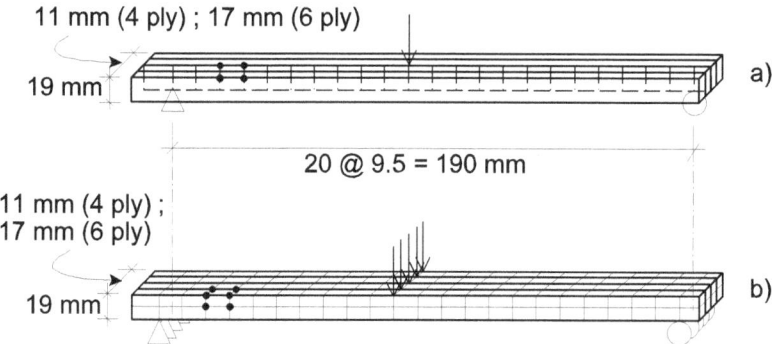

Figure 6.18 - Finite Element Mesh for Bending Configuration a) 2 Dimensions b) 3 Dimensions

Table 6.7 - Finite Element Mesh Properties of Angle Ply Laminate Bending Analyses

Finite Element Mesh Properties	2 Dimensional Model				3 Dimensional Model			
	0	[±15]$_s$	[±30]$_s$	90	0	[±15]$_s$	[±30]$_s$	90
Number of nodal points	69				525	345	345	525
Number of degrees of freedom	138 44				1575	1035	1035	1575
Number of elements					288	176	176	288

The grading of the mesh, in this case, was partially dictated by computer capacity. Simply, the 3 dimensional model was not able to process the problem using the same element size as for the uniaxial tests (4.5 x 4.5 mm^2 in the x - y coordinate system). Therefore, to weigh the effects of modeling error, a parametric study was carried out.

Parametric study:
In order to validate the element size chosen, two different element sizes were investigated using the 3 dimensional program for the above defined bending analysis. The results of an isotropic analysis (one element through the thickness to accommodate processor limitations) using the x - y coordinate mesh outlined in Figure 6.18 was compared to that of a mesh doubly refined in length and height. The input parameters of the analysis were the mean values in Table 6.5 (grain angle of 0°). The load - mid-span displacement curves for both results is shown in Figure 6.19.

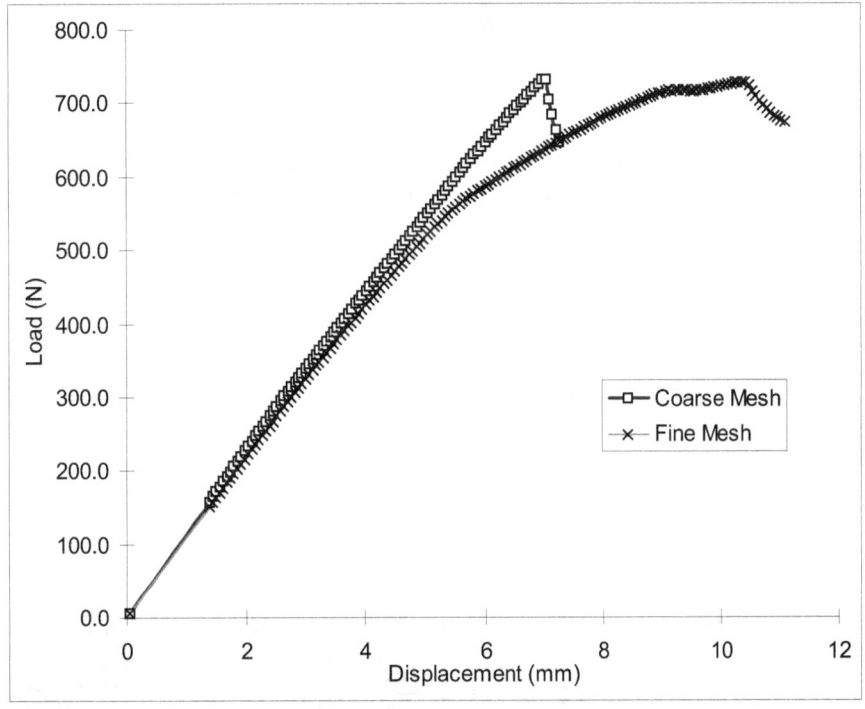

Figure 6.19 - Comparison of Coarse and Fine Mesh for Bending Analysis

Although both curves are clearly distinct, it is important to note that both initial stiffness and ultimate load are approximately equal. However, the fine mesh produces a more ductile curve, due to the higher number of Gaussian points present. For this one layer analysis, this is a consequence of the stresses being more easily redistributed with a finer mesh. It remains to be seen if this feature is helpful in accurately predicting the experimental findings. For the time being, the coarse mesh is accepted and its robustness will be borne out through comparison with experimental data.

The input properties for the academic laminates differed from that outlined in Table 6.5 only in the values used for tensile capacities as a result of size effect. The tension perpendicular-to-grain values, varying with each Gaussian point, were adjusted using Equation 5.5 as described in Section 6.3.1.1. In so doing, it was assumed that at each Gaussian point the stresses were, on an infinitesimal scale, uniformly distributed about the point. The tension parallel-to-grain strength values were, on the other hand, adjusted for each strand. To do this, a load configuration effect was implemented.

Based on similar principles as for size effect, load configuration effect is the result that brittle strength is dependant on the proportion of material that is highly stressed. For example, a pure tensile member, which has its entire volume highly stressed, has a higher probability of a critical flaw and hence a lower strength than does a bending member of the same size which has less of its volume highly stressed. The relationship between tensile and bending failure stress is derived as follows (Clouston, 1995):

We equate the probability of failure in bending to the probability of failure in pure tension. Given Equation 5.3

$$1 - e^{-\frac{1}{V_o} \int_{V_1} \left[\frac{\tau_1}{m} \right]^{\beta} dV_1} = 1 - e^{-\frac{1}{V_o} \int_{V_2} \left[\frac{\tau_2}{m} \right]^{\beta} dV_2}$$

Where we let: $\tau_1 = \sigma_{xb}$ = longitudinal tension stress in bending
$\tau_2 = \sigma_{xt}$ = longitudinal tension stress in pure tension

Considering the 3 point bending specimen first:

If $\sigma_{max(b)}$ = maximum longitudinal tension stress in elastic bending
M = bending moment along longitudinal axis
M_{max} = maximum bending moment

then,

$$\sigma_{max(b)} = \frac{M_{max}}{S} = \frac{FL}{4} \left(\frac{6}{bd^2} \right) \tag{6.1}$$

Rearranging and simplifying, the expression for maximum elastic load, F is

$$F = \frac{2bd^2}{3L} \sigma_{max(b)} \tag{6.2}$$

The longitudinal tension stress in bending at any point in the x - y plane is known from linear elastic beam theory as

$$\sigma_{xb}(x,y) = \frac{M(x) \cdot y}{I} = \left(\frac{F \cdot x}{2}\right)\left(\frac{12}{bd^3}\right) \cdot y \qquad (6.3)$$

Substituting F from Equation 6.2 and simplifying, we have

$$\sigma_{xb}(x,y) = \frac{4\sigma_{max(b)} \cdot x \cdot y}{L \cdot d} \qquad (6.4)$$

The probability of failure, P_{fb} in bending is now written as

$$P_{fb} = 1 - e^{-\frac{1}{V_o} \int_V \left(\frac{4\sigma_{max}}{Ldm} \cdot xy\right)^{\beta} dV} \qquad (6.5)$$

which resolves as follows

$$P_{fb} = 1 - e^{-\frac{1}{V_o} \int_V \left(\frac{4\sigma_{max}}{Ldm} \cdot xy\right)^{\beta} dV}$$

$$= 1 - e^{-\frac{2}{V_o} \int_{-b/2}^{b/2} \int_{-d/2}^{0} \int_{0}^{L/2} \left(\frac{4\sigma_{max}}{Ldm} \cdot xy\right)^{\beta} dx \cdot dy \cdot dz}$$

$$= 1 - e^{-\frac{2}{V_o} \cdot b \left(\frac{4\sigma_{max}}{Ldm}\right)^{\beta} \int_{-d/2}^{0} \int_{0}^{L/2} (xy)^{\beta} dx \cdot dy}$$

$$= 1 - e^{-\frac{2b}{V_o} \left(\frac{4\sigma_{max}}{Ldm}\right)^{\beta} \frac{(xy)^{\beta+1}}{(\beta+1)^2} \Big|_{0}^{L/2} \Big|_{-d/2}^{0}}$$

$$= 1 - e^{-\frac{2b}{V_o} \left(\frac{4\sigma_{max}}{Ldm}\right)^{\beta} \frac{\left(\frac{Ld}{4}\right)^{\beta+1}}{(\beta+1)^2}}$$

Finally,

$$P_{fb} = 1 - e^{-\frac{2bLd}{V_o 4} \left(\frac{\sigma_{max}}{m}\right)^{\beta} \frac{1}{(\beta+1)^2}}$$

$$= 1 - e^{-\frac{V}{2V_o} \left(\frac{\sigma_{max}}{m}\right)^{\beta} \frac{1}{(\beta+1)^2}}$$

Considering the pure tension case,

$$P_{ft} = 1 - e^{\frac{-V_t}{V_o}\left(\frac{\sigma_{max(t)}}{m}\right)^{\beta}}$$

(6.6)

Equating the two probabilities

$$\frac{-V_t}{V_o}\left(\frac{\sigma_{max(t)}}{m}\right)^{\beta} = \frac{-V_b}{V_o}\left(\frac{\sigma_{max(b)}}{m}\right)^{\beta}\frac{1}{2\cdot(\beta+1)^2}$$

(6.7)

whereupon we obtain the relationship

$$\frac{\sigma_{max(t)}}{\sigma_{max(b)}} = \left(\frac{V_b}{2V_t(\beta+1)^2}\right)^{\frac{1}{\beta}}$$

(6.8)

The tension parallel-to-grain strength for the 0° laminate (for example) is then adjusted as follows:

Given:
 $\beta = 4.2$
 $X_{t1} = \sigma_{max(t)} = 68.77$ MPa
 $V_t = 18.8 \times 3.2 \times 50.3 = 3026.0$ mm^3
 V_b (0°laminate) $= 17.4/6 \times 190 \times 19 = 10{,}469$ mm^3
 then

$$X_{t2} = X_{t1}\cdot\frac{1}{\left(\dfrac{V_b}{2V_t(\beta+1)^2}\right)^{\frac{1}{\beta}}} = 68.8\cdot\frac{1}{\left(\dfrac{10469}{2\cdot 3026(4.2+1)^2}\right)^{0.24}} = 132.7\,\mathrm{MPa}$$

The coefficient of variation is assumed constant so that the standard deviation becomes
132.7 MPa x 0.2667 = 35.4 MPa

It is noted here that Equation 6.8 is valid for the current laminate under consideration (ie. with a depth equal to that of one strand). If laminates with several strands through the depth were considered, the formulation would need to be reassessed. For a laminate with 4 strands in depth, for example, the integration in Equation 6.5 would be carried out over the y coordinates of the strand in question.

This computation was carried out for all academic laminates and the final values used for input are tabulated in Table 6.8. The tension perpendicular-to-grain strengths were adjusted in the usual manner using Equation 5.5.

Table 6.8 - Tensile Strength Properties used for 3 Point Bending Analyses

Strength		2 Dimensional		3 Dimensional	
		Mean	St. Dev.	Mean	St. Dev.
Parallel-to-grain Tension (X_t)	[0°] (MPa)	132.7	35.4	132.7	35.4
	[±15]$_s$ (MPa)	134.4	35.8	134.4	35.8
	[±30]$_s$ (MPa)	136.2	36.3	136.2	36.3
	[90°] (MPa)	132.8	35.4	132.8	35.4
Perpendicular-to-grain Tension (Y_t)	[0°] (MPa)	5.18	0.95	5.75	1.05
	[±15]$_s$ (MPa)	5.23	0.96	5.80	1.06
	[±30]$_s$ (MPa)	5.27	0.96	5.85	1.07
	[90°] (MPa)	5.19	0.95	5.75	1.05

The stresses are calculated and monitored at the Gaussian points for yielding or failure. However, failure naturally ensues at the extreme fiber in bending , where, in actuality, the longitudinal stresses are highest. To reflect this, the longitudinal stresses are extrapolated linearly from the Gaussian points to the outer edge of the beam prior to being checked in the failure / yield criterion.

Referencing Table 6.5, the initial stiffness property of a strand is different depending on tensile or compressive stresses as well as material direction (parallel or perpendicular-to-grain). As previously noted, the bending analysis produces multiaxial states of stress with all combinations of tension, compression, parallel-to-grain, or perpendicular-to-grain stress states. The stress state of each point is not known, however, until the first load (or displacement) is applied. Thus, it was necessary to devise a method to designate the appropriate initial stiffness corresponding to the unknown stress state of a Gaussian point. This was managed by first establishing the stress state for all Gaussian points (in the first iteration of the first increment) using default stiffnesses (tension assumed) with a small displacement increment. Figure 6.20 shows an extreme exploded view of the origin of the load displacement curves for [±15]$_s$ angle ply laminates in bending. Being linear elastic in this small range, the problem converges on the first iteration. Thus, in the first iteration of the second increment, the stiffnesses are updated to reflect the stress sign (i.e. if negative, then compressive; if positive, then tensile). The program then proceeds with larger displacement increments based on those stiffness values.

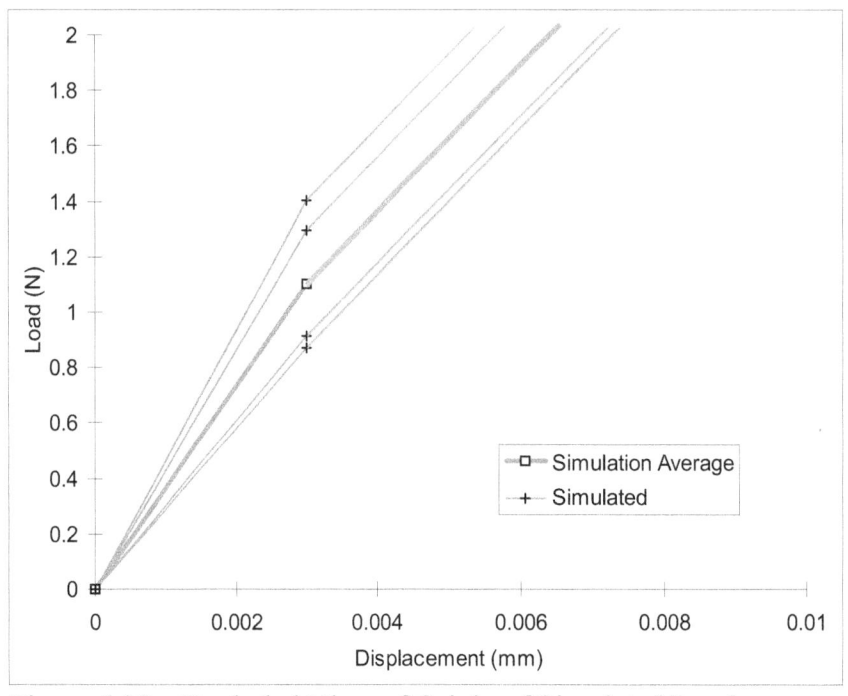

Figure 6.20 - Exploded View of Origin of Simulated Load
Displacement Curves for [±15]ₛ Angle Ply
Laminates

6.4.2 Experimental Tests

Laboratory tests were conducted on the 4 categories of academic laminates: $[0°]$, $[\pm15]_s$, $[\pm30]_s$, and $[90°]$. All tests were performed using a 133 kN capacity Sintech model 30/D universal testing machine. The 3 point bending setup was arranged as illustrated in Figure 6.21. In all cases the clear span was 190 mm. Load was applied under displacement control mode at a constant rate of 1.02 mm/min. to achieve failure in, on average, 5 minutes. Specimen cross sectional dimensions were nominally 19 in depth and 17 mm in width for the $[0°]$ and $[90°]$laminates - 11 mm in width for the $[\pm15]_s$ and $[\pm30]_s$ angle-ply laminates. Dimensions were measured at ends and mid-span using digital calipers and were found to be reasonably consistent with a maximum coefficient of variation of 1.8 percent for thickness and 0.4 percent for depth.

Figure 6.21 - 3 Point Bending Test Setup

6.4.3 Results

The simulated and experimental ultimate loads for the 3 point bending analyses are plotted in cumulative probability form in Figures 6.22 and 6.23 for the 2 dimensional and 3 dimensional setups, respectively. The descriptive statistics are outlined in Table 6.9.

Table 6.9 - Experimental and Simulated Data for [0°], [±15]$_s$, [±30]$_s$ and [90°] Laminates in 3 Point Bending

Config.	Statistic	Experiment	Simulation (Count = 500)			
		{Count}	2 d	(% error)	3 d	(% error)
[0°]	Initial Stiffness Mean (MPa)	12735.7 {20}	8456.4	33.6	9999.4	21.5
	Initial Stiffness COV (%)	6.9	7.5	8.7	8.2	18.8
	Ultimate Load Mean (N)	2967.5	2204.6	25.7	2622.0	11.6
	Ultimate Load COV (%)	8.6	5.4	37.2	6.7	22.1
[±15]$_s$	Initial Stiffness Mean (MPa)	9023.3 {20}	9172.5	1.6	8473.0	6.1
	Initial Stiffness COV (%)	17.2	12.6	26.7	12.0	30.2
	Ultimate Load Mean (N)	825.6	757.6	8.2	849.8	2.9
	Ultimate Load COV (%)	16.0	9.0	43.8	11.5	28.1
[±30]$_s$	Initial Stiffness Mean (MPa)	4292.2 {23}	4870.8	13.5	4925.9	14.8
	Initial Stiffness COV (%)	5.7	11.1	94.7	11.6	103.5
	Ultimate Load Mean (N)	444.8	301.8	32.1	401.4	9.8
	Ultimate Load COV (%)	8.9	7.3	18.0	8.7	2.2
[90°]	Initial Stiffness Mean (MPa)	839.5 {19}	143.5	82.9	187.9	89.9
	Initial Stiffness COV (%)	15.1	16.1	6.6	6.5	63.8
	Ultimate Load Mean (N)	98.8	159.0	60.9	187.6	89.9
	Ultimate Load COV (%)	13.0	7.4	43.1	4.7	63.8

The 2 dimensional program is clearly less accurate in predicting the experimental findings than its 3 dimensional counterpart. With regards to ultimate load prediction, in each case, with the exception of the 90° laminate, the 3 dimensional program is more accurate in describing the entire sample set (i.e. both mean and variability). It is speculated that the ability of the 3 dimensional model to account for interlaminar shear stresses leads to a more realistic and therefore more accurate analysis. Although the through thickness stresses are not considered directly in the yield / failure criterion, the mere presence of these stresses in the calculations provide a more precise computation of the in-plane stresses. It is fair to say that the present stochastic bending analysis, with different stiffnesses for each ply, requires a 3 dimensional analysis to account for the complexities.

This said, the 3 dimensional program requires approximately 8 times more computer space and time (reference Table 6.7). The 2 dimensional analysis, as an alternative, provides a rough approximation for the average ultimate load (maximum error of 32.1 percent) for the $0°$, $[\pm15]_s$ and $[\pm30]_s$ laminates.

With regards to predicting initial stiffness of the curves, both programs produce reasonable results. Again, with the exception of the $90°$ laminate, the worst case result is that of the $0°$ laminate with an error of the mean of 33.6 percent . It is speculated that this relatively high error may be a result of local crushing at the node points which was prevalent for the $0°$ laminates in the experiments. The dominant compressive stresses may have contributed to more Gaussian points with compressive elasticity resulting in an overall more compliant laminate stiffness.

Ultimate load for the $0°$ laminate is under-predicted for both the 2 and 3 dimensional model (25.7 % error for 2 D. and 11.6 % error for 3 D). Due to factors of scale, this fact is emphasized in Figures 6.22 and 6.23. It is speculated that this inaccuracy is a result of the programs' inability to simulate local crushing at the load head to the extent seen in the experimental specimens. Referencing Figure 2.2, the interaction parameter, F_{12}, was determined in the second quadrant of the strength envelope, where the stress state is compression parallel-to-grain and tension perpendicular-to-grain. For the $0°$ laminate bending case, the third quadrant (compression parallel-to-grain and compression perpendicular-to-grain) governs initially. It may be that the Tsai Wu strength envelope is overly restrictive in this quadrant. A full investigation could be carried out in a future study to assess the accuracy of this theory in all quadrants. In the same respect, alternative strength criterions for wood strand composites could also be surveyed.

The $90°$ laminates are poorly estimated on all accounts. The cause of this may be experimental in nature. The $90°$ laminates were very fragile and susceptible to cupping. It is possible that the experimental specimens (for tensile tests as well as bending) contained micro-cracks or dimensional irregularities which negatively impacted the results.

The load displacement curves for the entire range of experimental and simulated results are given in Figures 6.24 - 6.31. A total of 500 replications were run for each configuration; however, as before, only the stiffest, least stiff, strongest and weakest as well as curve averages for both sets are shown for clarity.

All simulations, save the $90°$ laminates, seem to replicate the experimental data well. The ductile behaviour for the $0°$ and $[\pm15]_s$ laminates is obvious, as is the onset of ultimate failure by abrupt brittle behaviour. The program, in its present state, however, is unable to capture the experimental feature of recovery where, upon abrupt local failure, the stresses redistribute to surpass the previous peak. This is likely a consequence of the slow release of stresses upon brittle failure (per Table 4.2) to assure convergence of the system. The stresses do indeed redistribute numerically, but more gradually.

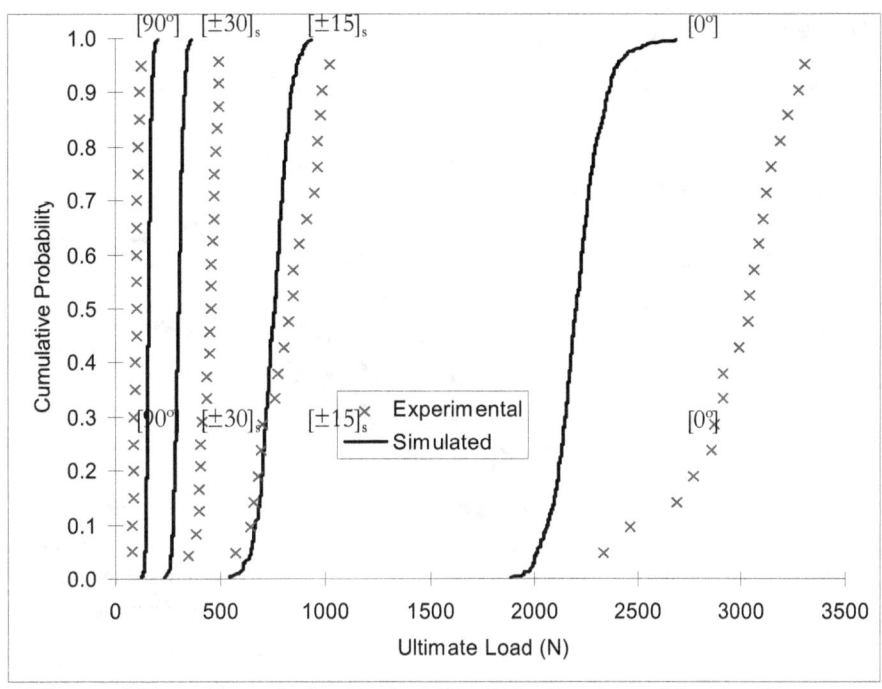

Figure 6.22 -   Cumulative Probability Distribution of Academic
                Laminates in 3 Point Bending (2 - Dimensional Model)

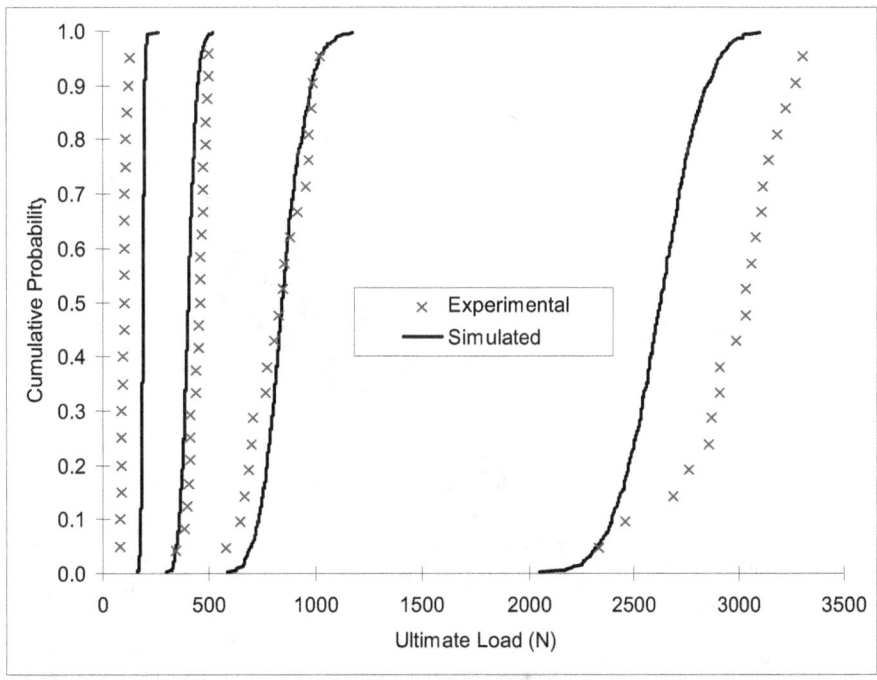

Figure 6.23 -   Cumulative Probability Distribution of Academic
                Laminates in 3 Point Bending (3 - Dimensional Model)

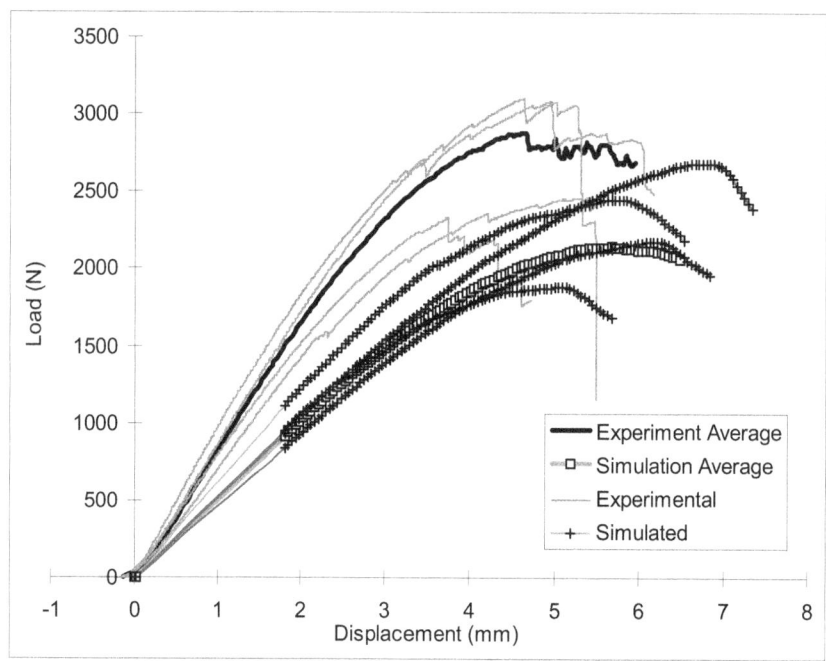

Figure 6.24 - Load-Displacement Curves of [0°] Laminate in Bending (using 2 - Dimensional Model)

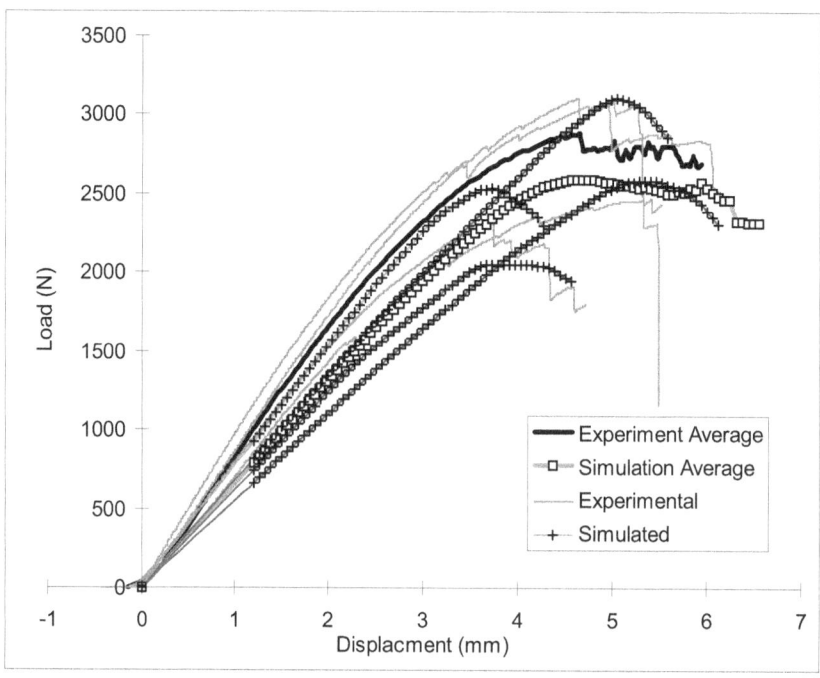

Figure 6.25 - Load-Displacement Curves of [0°] Laminate in Bending (using 3 - Dimensional Model)

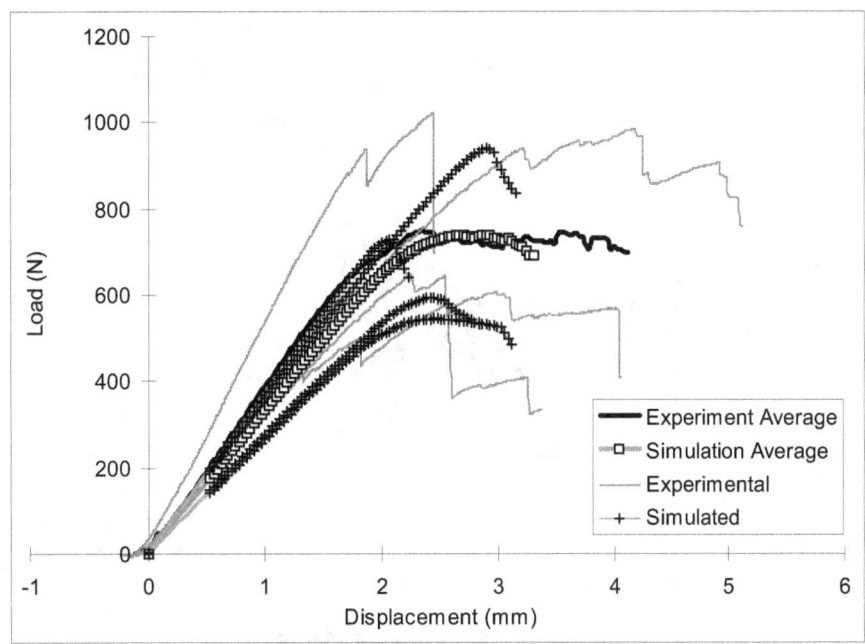

Figure 6.26 - Load-Displacement Curves of [±15]_s Angle Ply Laminate in Bending (using 2 - Dimensional Model)

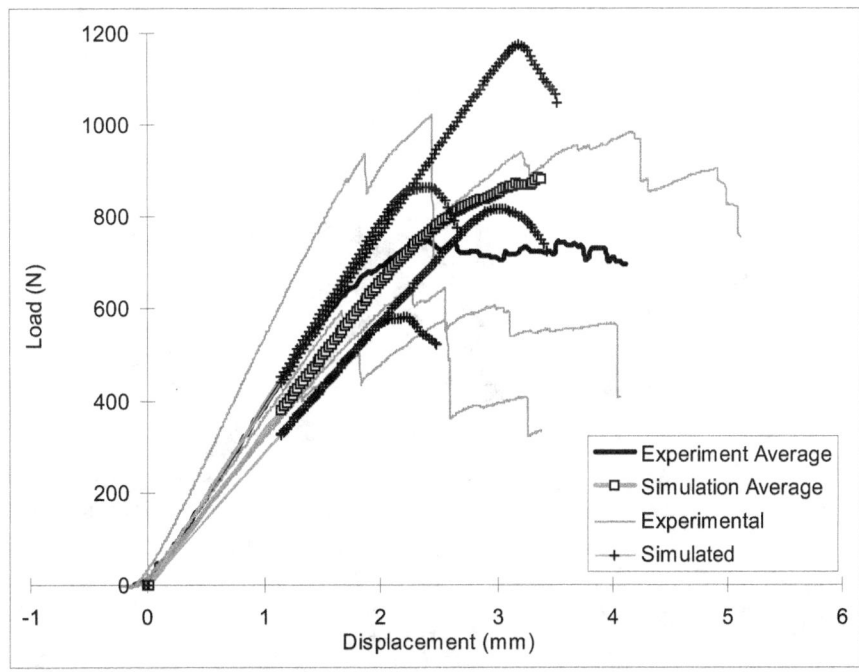

Figure 6.27 - Load-Displacement Curves of [±15]_s Angle Ply Laminate in Bending (using 3 - Dimensional Model)

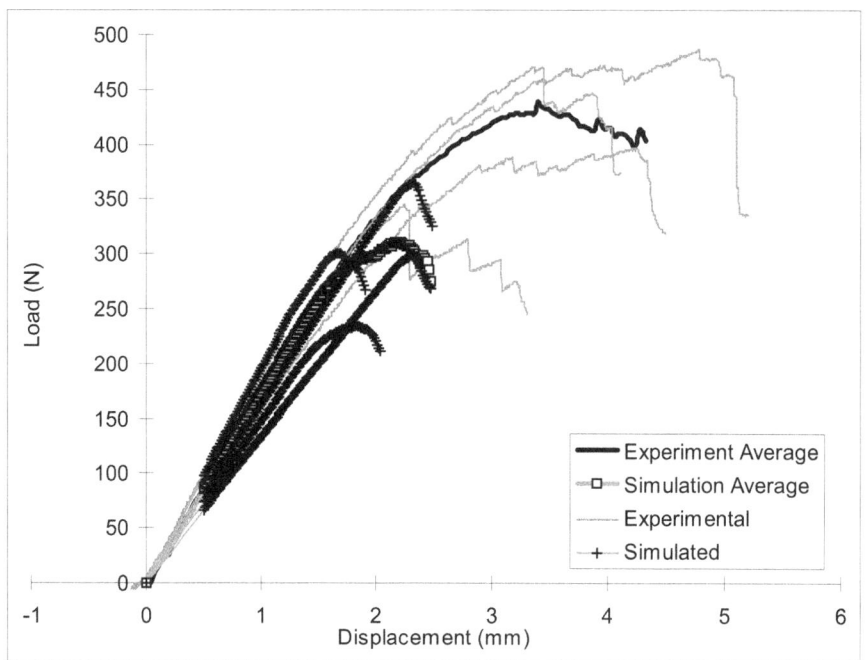

**Figure 6.28 -   Load-Displacement Curves of [±30]ₛ Angle Ply
Laminate in Bending (using 2 - Dimensional Model)**

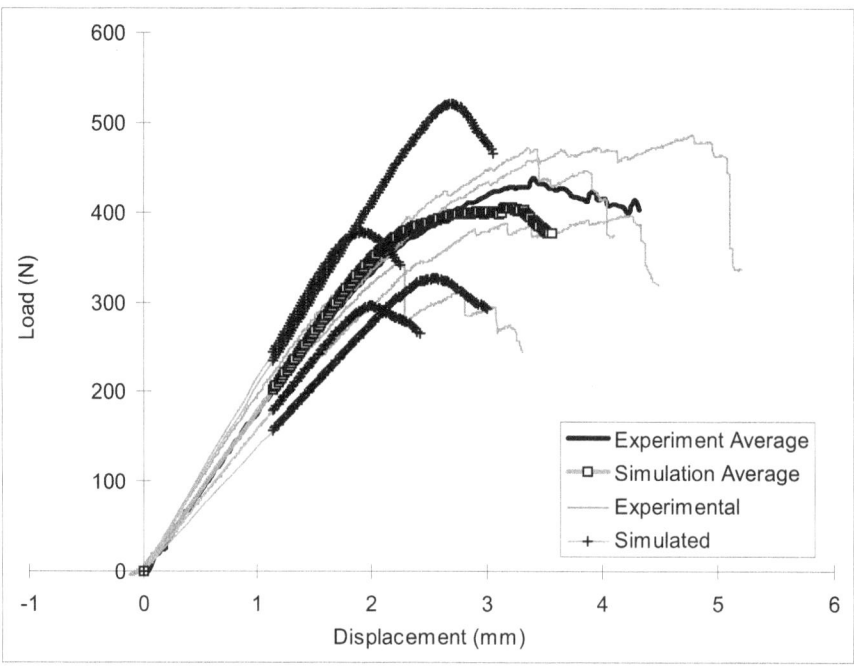

**Figure 6.29 -   Load-Displacement Curves of [±30]ₛ Angle Ply
Laminate in Bending (using 3 - Dimensional Model)**

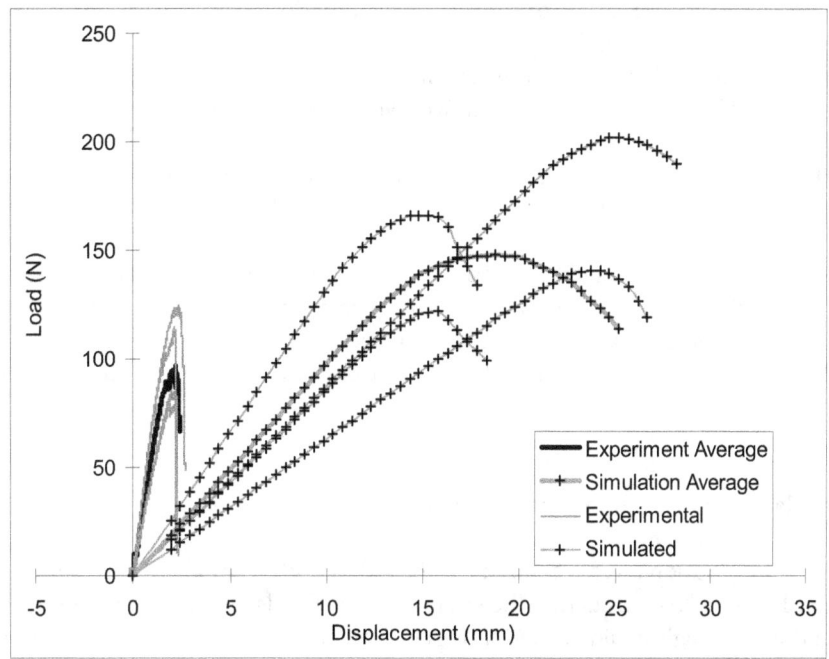

Figure 6.30 -   Load-Displacement Curves of [90°] Laminate in Bending (using 2 - Dimensional Model)

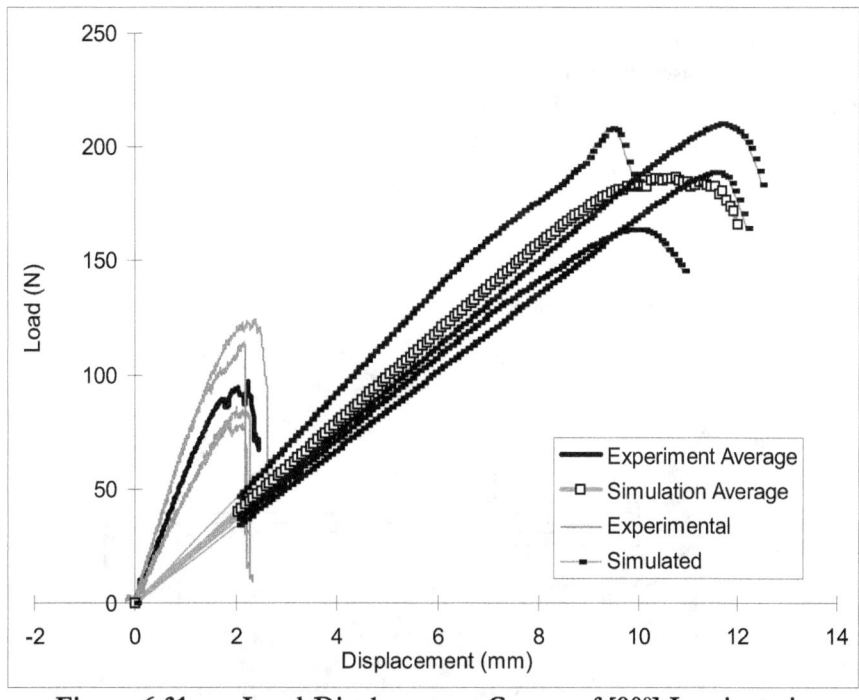

Figure 6.31 -   Load-Displacement Curves of [90°] Laminate in Bending (using 3 - Dimensional Model)

6.5 Parallam® Investigation

As a preliminary assessment of the model, a comparison of simulated and experimental data for Parallam® was conducted. It should be understood that at this stage in the model development, complete accuracy is not expected. The model is not yet formulated to address all the complexities of the spatial relationship between individual wood elements in wood composites. It was felt that a general comparison at this stage would simply be useful in providing insight into ultimately simulating commercial products. This section addresses the current method involved in modeling grain angle deviations in Parallam®, as well as a comparison between simulated and experimental findings for tensile and bending configurations.

The 2 dimensional program, as a consequence of the assumptions made for the classical lamination theory, is restricted to symmetric laminate analysis. The 3 dimensional program, on the other hand, is configured to analyze laminates with any stacking sequence. Because the lay-up of strands - and therefore grain angle - is known to be random for Parallam®, the following investigation focuses solely on the 3 dimensional model.

6.5.1 Simulation of Parallam® Strand Lay-up

Special preparations were required for the 3 dimensional model to accommodate the complex strand lay up of Parallam®. Considering the strands are randomly aligned with the longitudinal axis, it was decided that the grain angle could be randomly generated based on a uniform distribution of approximate observed measurements. These grain angle characteristics were physically measured as outlined below:

Ten individual boards of Coastal Douglas-fir 2.0E (ie. 2.0x10^6 psi or 13.8 GPa modulus of elasticity) Parallam® were used for both tensile specimens as well as grain angle distribution measurements. The 3 dimensional mapping region lay directly adjacent to where the specimens were cut from the board, as shown in Figure 6.32. Visual measurements of maximum grain angle were estimated over 25 x 16 mm^2 elements along the length of the board. Maximum grain angle was used because it was assumed that for the majority of cases, the maximum deviation of grain would govern strength (for example, a knot within an element would be the weakest point and would constitute a 90° grain angle). A layer of 3mm was then planed from the board and measurements were taken from the new surface. This was done for 4 successive layers to produce a 12 mm through-thickness profile.

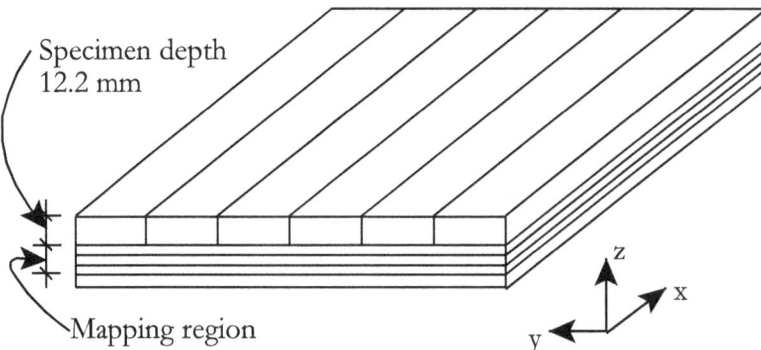

Figure 6.32 -   Cutting and Grain Angle Mapping Template for Parallam®

The data was then interpreted into a frequency distribution as seen graphically in Figure 6.33. From this data, a uniform distribution was created with equal probability of having the same angle as positive or negative. This uniform distribution formed the basis of random angle generation within the program.

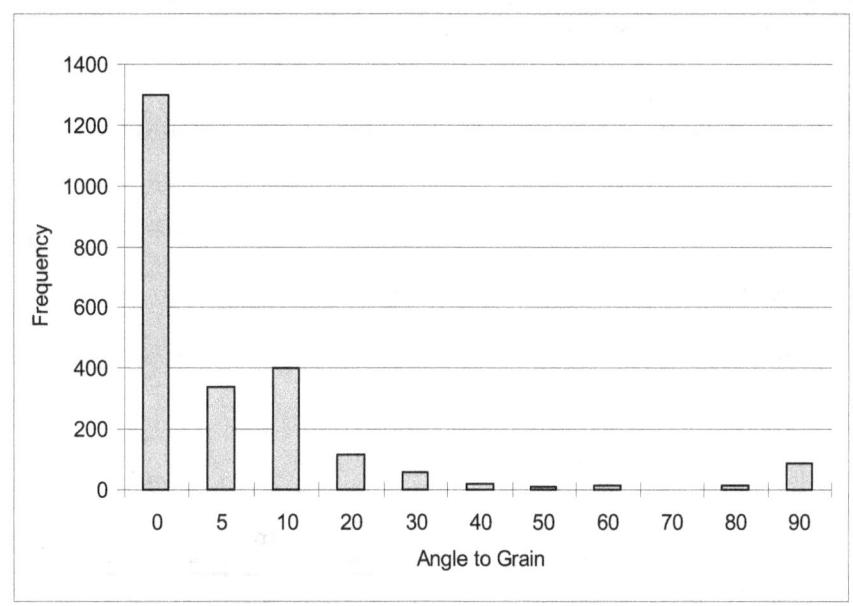

Figure 6.33 - Frequency Distribution of Measured Parallam® Grain Angle

6.5.2 Parallam® Tension Comparison

6.5.2.1 Numerical Analysis

Having completed this addition to the program, the input file was then formulated for the tensile specimen. The finite element mesh assumes a uniform stress field between the grips and is illustrated in Figure 6.34. This mesh considers the strand lay up to be substantially simplified in comparison to that of Parallam®. The model assumes a total of 4 strands through the thickness per specimen. This is not wholly realistic as the strands in Parallam® are not planar or continuous and tend to meander in three directions. Furthermore, voids are prevalent throughout the material. It is possible that future versions of the program could incorporate this geometric variability.

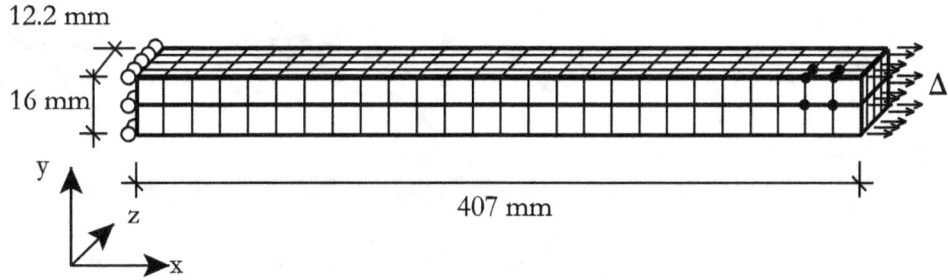

Figure 6.34 - Finite Element Mesh for Parallam® in Tension

To account to some degree for the geometric variability, the tension parallel and perpendicular-to-grain strengths were assumed to vary from integration point to integration point. However, because the dimension in the through-thickness (z) direction was very narrow (3.1 mm), the adjacent points in the z direction were given the same tension parallel-to-grain properties. These values were adjusted in the usual manner (described in Section 6.3.1.1) for size effect.

The strand properties are that of the heartwood Douglas-Fir database established in Chapter 5. It is noted, however, that Parallam® may contain both of heartwood and sapwood.

6.5.2.2 Experimental Tests

The rectangular specimens were shaped into a 'dogbone' profile to initiate failure within the gauge length. Without this special profile, failure would consistently occur in or near the grips due to stress concentrations from grip crushing. The dimensions of the specimens are given in Figure 6.35.

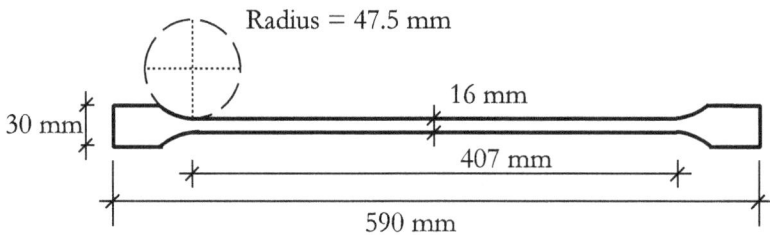

Figure 6.35 - Parallam® Tensile Specimen Dimensions

A total of 56 specimens were tested using a 222 kN MTS machine with a mechanical wedge-type grip system. The loading rate was 1.27mm/min. An extensometer with a 25.4 mm gauge length was employed to determine tensile MOE at the mid-span of the specimen. Typical failure occurred near mid-span in a splintered manner as depicted in Figure 6.36.

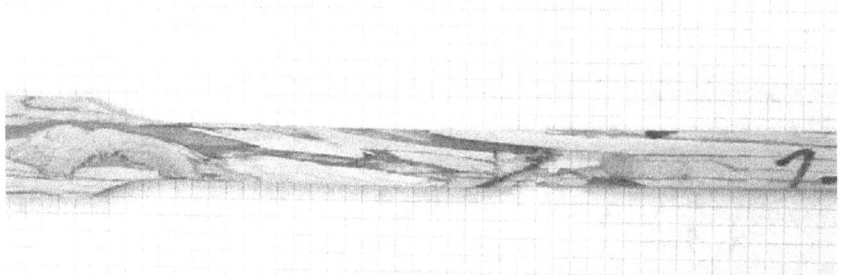

Figure 6.36 - Tensile Failure of Parallam® Specimen

6.5.2.3 Results

The simulated data is compared with the experimental results in Table 6.10 and Figure 6.37. In this preliminary investigation, the elastic modulus was under-predicted which may be explained by the fact that each strand was allocated an angle to grain without variation. In actuality, the grain angle changes more frequently throughout the specimen. A more accurate depiction could be to vary the grain angle with each element.

The mean strength is substantially over-predicted. This is explained by the fact that the model, in its current state, is not formulated to manage all of the complexities of Parallam®. As previously noted, voids were not accounted for, which could lead to a significant over-estimation of strength. Moreover, the input properties were solely for heartwood Douglas-fir, although Parallam® is comprised of both heartwood and the weaker material sapwood.

Table 6.10 - Experimental and Simulated Data for Parallam® in Tension

Statistic	Experiment (count = 56)	Simulation (count = 500)	Percent Error (%)
Elastic Modulus Mean (MPa)	14,222.9	11,945.1	16.0
Elastic Modulus COV (%)	26.8	25.2	6.0
Strength Mean (MPa)	33.6	62.6	86.3
Strength COV (%)	25.2	20.4	19.1

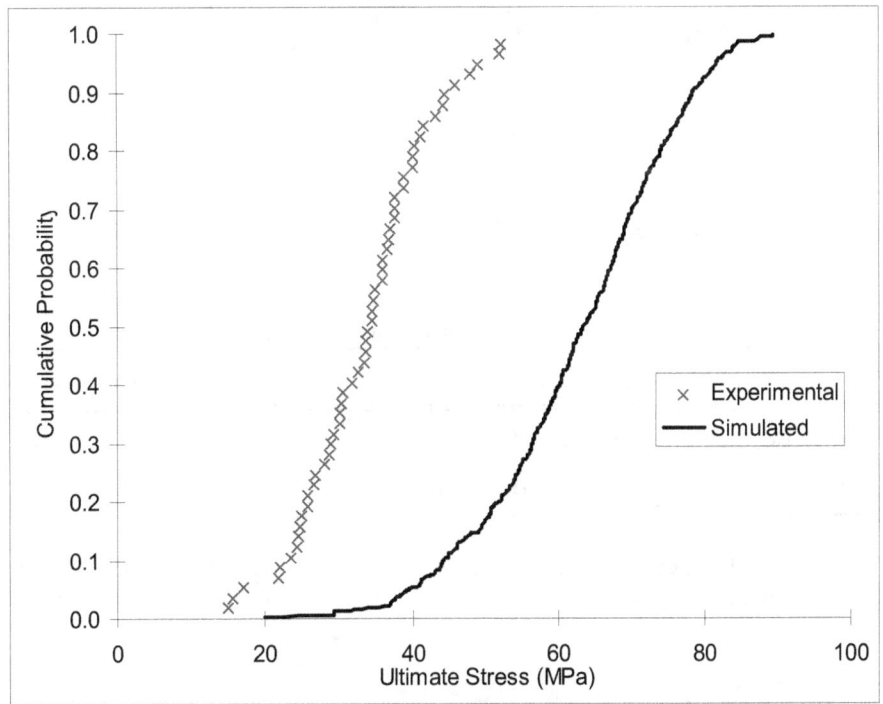

Figure 6.37 - Cumulative Probability Distribution of Parallam® in Tension (3 - Dimensional Model)

6.5.3 Parallam® Bending Comparison

A bending comparison was performed between simulated and experimental data for Parallam®, as was done for tension. It is again noted that this investigation is simply intended as a primary check. Modifications to the model to consider void volume, for example, should be further investigated.

6.5.3.1 Numerical Analyses

Figure 6.38 depicts the specimen geometry and finite element mesh for the bending analysis. The material is assumed to be comprised of 4 separate strands through the thickness each with different stiffness and strength properties. As before, for Parallam® subjected to tensile loading, the tensile strengths (both parallel and perpendicular-to-grain) were assumed to vary between Gaussian points to better approximate the random nature of the strands. Treating the stress distribution about each Gaussian point as uniformly distributed, the tensile strengths are adjusted for size effect using Equation 5.5. A summary of adjusted tensile strengths is provided in Table 6.11. All other input properties are taken from the acquired database for the Douglas-fir heartwood strands and are outlined in Table 6.5.

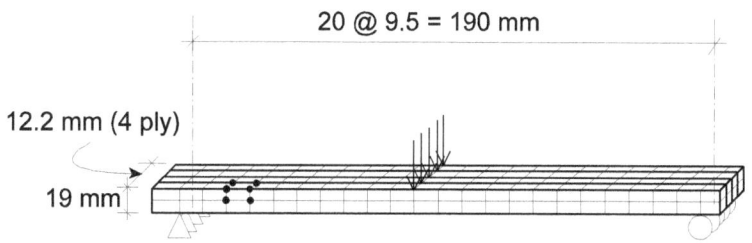

Figure 6.38 - Finite Element Mesh for Parallam® in 3 Point Bending

Table 6.11 - Tensile Input Properties for Parallam® Bending Analysis

Tensile Property	Mean	Standard Deviation
Parallel-to-grain Strength (X_t) (MPa)	171.27	45.7
Perpendicular-to-grain Strength (Y_t) (MPa)	4.62	0.85

The grain angle for each strand was randomly generated in accordance with the uniform distribution shown in Figure 6.33.

6.5.3.2   Experimental Tests

Bending tests were conducted using a Sintech model 30/D universal testing machine with a capacity of 133 kN - the same as for the academic material in section 6.4.2. The setup is illustrated in Figure 6.39 for a typical Parallam® specimen. Load was applied with a uniform loading rate of 0.9mm/minute under displacement control mode. Specimen cross sectional dimensions were measured at ends and mid-span and were found to have high consistency with a maximum coefficient of variation of 1.7 percent for thickness and 0.7 percent for depth.

Figure 6.39 -   3 Point Bending Test Setup for Parallam

6.5.3.3   Results

A statistical summary of experimental and simulated data is outlined in Table 6.12 and visually represented in Figure 6.40. The entire range of load-displacement curves are shown in Figure 6.41. Considering the model is not regarded as complete at this stage, the results are quite reasonable. Ultimate load average is slightly over-predicted which is understandable given that voids are not considered and that the database comprised only of heartwood without the lower grade sapwood (which is present in equal proportions in Parallam®). The initial stiffness average is slightly under predicted, although referencing Figure 6.41, the range of results lay close to the experimental bounds.

The mode of failure is captured particularly well for Parallam®. In experiment it was observed that in some cases, the specimens tended to form compression wrinkles near the location of applied load prior to a splitting tensile failure at the bottom edge. This sequence of events was readily seen in both the experimental and simulated load - displacement curves. The curves follow a nonlinear path during predominant compressive failure and ultimately fail in a brittle manner due to predominant tensile stress failure.

The assumptions made in this analysis greatly simplify the geometric properties of Parallam®. A more rigorous study to ascertain and subsequently model the Parallam® strand layup should be the next step. From this preliminary analysis, however, it appears that the theory presented herein is appropriate and capable of ultimately predicting the mechanical behaviour of wood strand composites in general.

Table 6.12 - Experimental and Simulated Data for Parallam® in 3 Point Bending

Statistic	Experiment	Simulation (Count = 500)	
	{Count}	3 d	(% error)
Initial Stiffness Mean (MPa)	9239.4{41}	8050.8	12.9
Initial Stiffness COV (%)	12.3	12.7	3.3
Ultimate Load Mean (N)	1232.8	1317.9	6.9
Ultimate Load COV (%)	16.4	13.5	17.7

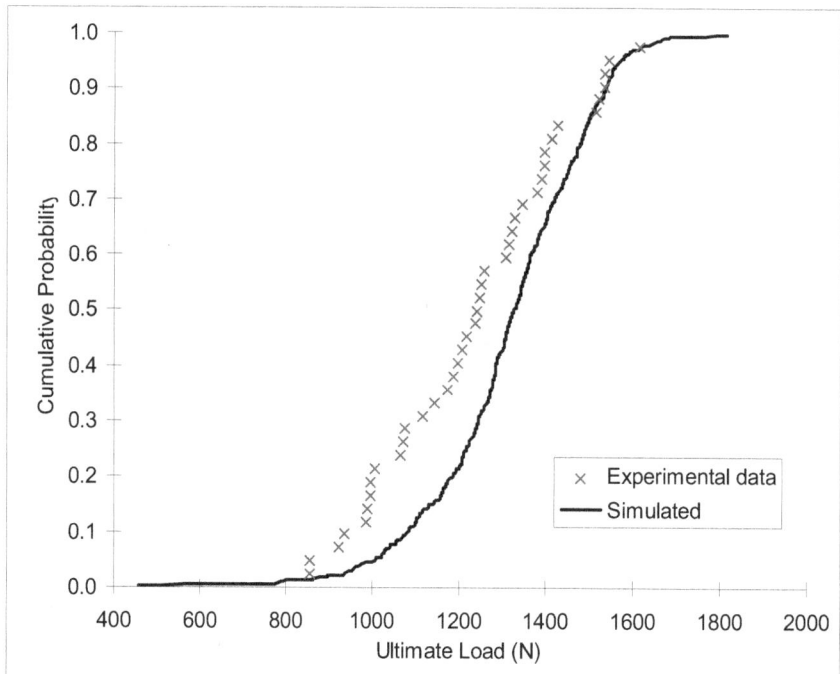

Figure 6.40 - **Cumulative Probability Distribution of Parallam® in 3 Point Bending (3 - Dimensional Model)**

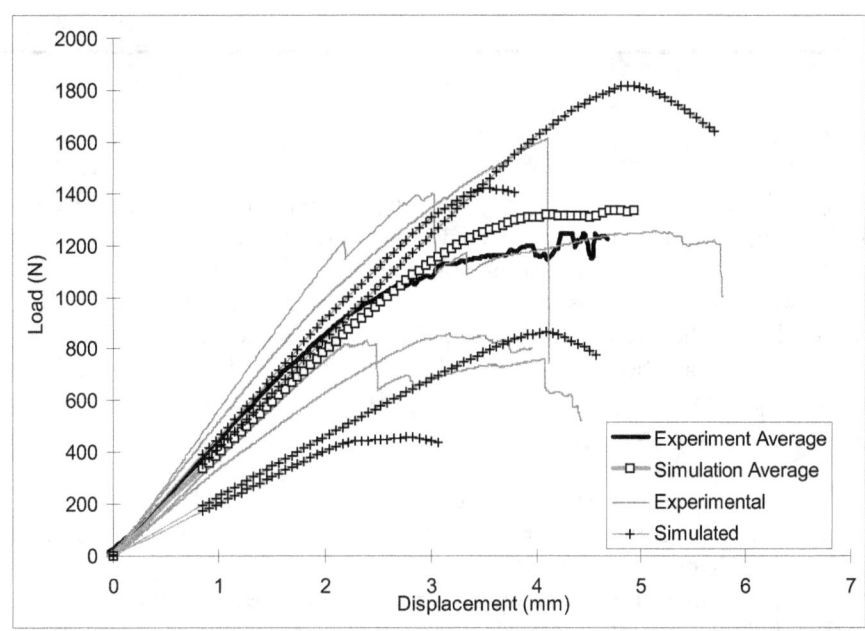

Figure 6.41 -   Load-Displacement Curves of Parallam® in 3 Point
Bending (using 3 - Dimensional Model)

7.1 SUMMARY

Despite the advantages of having a computational strength prediction model for wood composite research and design, very few studies have been devoted to this subject in the literature. Work that has been carried out has concentrated to a large degree on linear elastic constitutive theory. The physical nature, however, of the wood strand composites utilized in today's lumber products industry necessitate a more rigorous approach. In the present study, a continuum mechanics approach, considering elastic and inelastic behaviour up to and beyond failure, has been investigated. The primary objective of the present study was to develop a computational model to simulate the mechanical response of wood strand composites subjected to bending.

A finite element based program, COMAP (COmposite Analysis Program), was developed. Two versions were created. A 2 dimensional version was developed which utilizes classical lamination theory (CLT) to evaluate resultant forces for symmetric multilayered laminates. This version, although efficient in computational time and general usage, is restricted by the assumption of symmetry and the assumptions of the CLT. Notably, plane stress is assumed within each layer and thus stresses in the through-thickness direction are neglected. A 3 dimensional version was created to provide a means of assessment for nonsymmetrical laminates and also to gauge the influence of through - thickness stresses.

An orthotropic elastic - plastic - failure constitutive model was implemented into both versions of the program. The nonlinear formulation (beyond the elastic regime) was managed using classic orthotropic incremental plasticity theory. The constitutive model utilized the Tsai-Wu yield / failure criterion to describe the initial and subsequent yield surfaces assuming an associated flow rule. All stresses were monitored for each iteration of each increment of loading (or displacement) and were initially transformed into the principal material directions prior to use in the yield criterion. Ultimate failure was defined using ultimate principal strengths (experimentally determined) to establish an upper bound yield surface. Upon reaching the upper bound surface, an integration point either 1) failed in a brittle manner or 2) first strain - hardened and then eventually failed in either a ductile or brittle mode. The latter choice was decided based on whether the combination of stresses at the point at the time of failure were predominantly tensile or predominantly compressive.

Stiffness and strength properties of the material were input as random variables for both programs. This allowed for many replications of one configuration through Monte Carlo simulations to generate entire sample output. Moreover, the programs were formulated for stochastic analyses with varying strength and stiffness values within each member.

A strand property database of strength and stiffness for Douglas-fir heartwood was acquired which was necessary as input into the programs. Laminated boards were first fabricated from which specimens were cut and then prepared for experimental tests. For shear strength, modulus of ridgidity and the interaction parameter of the Tsai-Wu theory, a least square minimization of error between simulated and experimental compression strength of $[\pm 15]_s$ angle ply laminates was conducted. This calibration procedure provided the mean and standard deviation of the 3 variables.

To verify the performance of the programs, the solutions to various problems were compared to known analytical solutions and that of other software. No one other software package has all of the capabilities of COMAP and consequently model verification was handled systematically to first address plastic deformation accuracy, then to check the linear elastic orthotropic finite element formulation, and finally to confirm the stress and strain transformation process required for multilayered laminates.

The effectiveness of the programs to replicate experimental findings was demonstrated in the final chapter. The results of numerical simulations for angle ply laminates in tension, compression and bending compared favorably with experimental data. Furthermore, preliminary comparisons of tensile and bending results for Parallam® showed the proposed technique to have great potential to ultimately model commercial wood strand composites.

7.2   CONCLUDING REMARKS

The favorable agreement between the numerical simulations and experimental data demonstrate the accuracy of the present technique. The analyses conducted herein point to the fact that orthotropic plasticity theory in conjunction with a stochastic finite element approach appear to offer an elegant solution to predicting wood composite behaviour. Indeed, this technique is general enough to be used for a variety of laminated composites provided the experimental database for the layer component of the composite is available. A full account of required data input for the present programs is provided in Appendix C along with sample outputs. With respect to this, it is cautioned that the present academic form of the programs requires further detailed investigation prior to use - particularly with commercial products.

7.3   FUTURE RELATED RESEARCH

The subject offers many interesting areas for further research. The following discussion provides a few ideas for extension and improvement of the current model.

An obvious improvement at this stage would be to upgrade the program's ability to describe the physical structure of commercial wood composites. One basic feature lacking from the present model is the ability to capture the complex 3 dimensional staggered stacking sequence of the strands which is the nature of wood composites although not common with advanced aero-spatial type composites. Several studies have been devoted to understanding the complex spatial relationship between individual wood elements in wood composites (Dai and Steiner, 1993; Ellis et al., 1994; Oudjehane et al., 1998; Lu, C. 1999). To incorporate these techniques into the present formulation may prove to be a difficult, although rewarding, task. It must entail consideration of density variations, potential null volumes or voids as well as varying 3 dimensional geometry of elements.

Many smaller alterations to the program could be investigated. Given the stochastic nature of the model, one could make any presently deterministic parameter stochastic. For example, the grain angle, in reality, likely varies within, as well as between, strands. Given an accurate mathematical depiction of this variation, properly implemented into the program, the model's present accuracy to simulate the experimental data may be enhanced. Presently, the compression properties are correlated in generating the compressive stress - strain curves in the principal material directions. Correlation between all variables could be considered, for example, within a strand. Also, the technique is not limited to the Tsai-Wu criterion nor the assumptions made for the description of plastic flow. Other yield surfaces (possibly considering non-associated flow) may be more appropriate for other materials.

Further areas of interest would include extension of the model to study wood based composite panels or investigation of nonlinear viscoelastic behaviour of wood composites. There is much to be learned of creep and creep rupture performance of wood composites under long term loading and varying temperature and humidity conditions. Also of strong interest would be adaptation of the model to investigate the interaction between mechanical connections and wood composites. The technique could potentially be extended to optimize strand layup for prescribed connections.

REFERENCES

ASTM Standard D143 1994. Standard Methods of Testing Small Clear Specimens of Timber. American Society for Testing and Materials, Philadelphia, Pa.

ASTM Standard D198 1994. Standard Methods of Static Tests of Timbers in Structural Sizes. American Society for Testing and Materials, Philadelphia, Pa.

ASTM D2016 1991. Standard Test Methods for Moisture Content of Wood. American Society for Testing and Materials, Philadelphia, Pa.

ASTM Standard D3518 / D3518M 1991. Standard Practice for In-Plane Shear Stress-Strain Response of Unidirectional Polymer Matrix Composites. American Society for Testing and Materials, Philadelphia, Pa.

Barrett, J.D.; Foschi, R.O.; Fox, S. P. 1975. Perpendicular-to-grain strength of Douglas-fir. Can. J. Civ. Eng., 2(1), pp. 50-57

Barrett, J.D.; Lam, F.; Lau, W. 1995. Size effects in visually graded softwood structural lumber. J. of Mat. in Civ. Eng., 7(1), pp. 19-30

Bodig, J.; Jayne, B.A. 1993. Mechanics of Wood and Wood Composites. Krieger Publishing Co. Malabar, Florida

Cha, J. K.; Pearson, R.G. 1994. Stress Analysis and Prediction in 3-Layer Laminated Veneer Lumber: Response to Crack and Grain Angle. Wood and Fiber Sc., 26(1), pp. 97-106

Chen, W.F.; Han, D.J. 1988. Plasticity for Structural Engineers. Springer-Verlag, New York Inc. New York

Clouston, P. 1995. The Tsai-Wu Strength Theory for Douglas-fir Laminated Veneer. Master of Applied Science Thesis, University of British Columbia, Vancouver, B.C. Canada

Clouston, P.; Lam, F.; Barrett, J.D. 1998. The Interaction Term of the Tsai-Wu Theory for Laminated Veneer, Journal of Materials in Civil Engineering, 10(2), pp.112-116

Clouston, P.; Lam, F; Barrett, J.D. 1998. Incorporating Size Effects in the Tsai-Wu Strength Theory for Douglas-fir Laminated Veneer. Wood Science and Technology, 32 (1), pp.215-226

Conners, T.E. 1989. Segmented Models for Stress-Strain Diagrams. Wood Sc. and Tech., 23(1), pp. 65-73

Cook, R.D.; Malkus, D.S.; Plesha, M.E. 1989. Concepts and Applications of Finite Element Analysis. John Wiley and Sons. New York

Cowin, S.C. 1979. On the Strength Anisotropy of Bone and Wood. ASME Trans., Journal of Appl. Mech., 46(4), pp. 832-837.

Dai, C.; Steiner, P. R. 1993. Spatial structure of wood composites in relation to processing and performance characteristics. Part 1. Rationale for model development. Wood Science & Technology, 28(1), pp. 45-51.

Ellis, S.; Dubois, J.; Avramidis, S. 1994. Determination of Parallam Macroporosity by Two Optical Techniques. Wood and Fiber Science, 26(1), pp.0-77

Foschi, R.O.and Yao, Z.C. 1986b. Another Look at Three Duration of Load Models. Procedings, IUFRO Wood Engineering Group Meeting, September; Florence, Italy: Volume 2, 19-1-1

Gibson, R. F. 1994. Principles of Composite Material Mechanics. McGraw-Hill Companies. New York

Goodman, J. R.; Bodig, J. 1971. Orthotropic Strength of Wood in Compression. Wood Science 4(2), pp. 83-94.

Hill, R. 1948. A Theory of the Yielding and Plastic Flow of Anisotropic Metals. Proceedings of the Royal Society, Series A, Vol. 193, pp. 281-297

Hull, D. 1981. An Introduction to Composite Materials. Cambridge University Press. Cambridge

Hunt, M. O.; Suddarth, S.K. 1974. Prediction of Elastic Constants of Particleboard. Forest Prod. J., 24(5), pp. 52-57

Jones, R,M. 1975. Mechanics of Composite Materials. McGraw-Hill, New York

Kobetz, R.W.; Krueger, G. P. 1976 Ultimate Strength Design of Reinforced Timber-Biaxial Stress Failure Criteria. Wood Science, 8(4), pp. 252-261

Lam, F., and Varoglu, E. 1990. Effect of Length on the Tensile Strength of Lumber. Forest Products J., 40(5), pp. 37-42

Liu, J.Y. 1984. Evaluation of the Tensor Polynomial Strength Theory for Wood. J. Comp. Mat., 18, pp. 216-225

Lu, C. 1999. Organization of Wood Elements in Partially Oriented Flakeboard Mats. Ph.D. Thesis, University of British Columbia, Vancouver, B.C. Canada

Madsen, B.; Buchanan, A.H. 1986. Size Effects in Timber Explained by a Modified Weakest-Link Theory. Can. J. Civ. Eng., 13(2), pp. 218-232

Madsen ,B. 1990. Size Effects in Defect Free Douglas-fir. Can. J. Civ. Eng., 17, pp. 238-242

Maghsood, J.; Scott, N.R.; Furry, R.B.; Sexsmith, R. 1973. Linear and Nonlinear Finite Element Analysis of Wood Structural Members. ASAE Trans. Paper No.70-922, pp. 490-496

Mase, G.E. 1970. Theory and Problems of Continuum Mechanics. Schaum's Outline Series, McGraw-Hill, Inc.

Nahas, M.N. 1986. Survey of Failure and Post-Failure Theories of Laminated Fiber-Reinforced Composites. J. Comp. Tech. and Res., 8(4), pp. 138-153

Narayanaswami, R.; Adelman, H. M. 1977. Evaluation of the Tensor Polynomial and Hoffman Strength Theories for Composite Materials. J. Comp. Mat.,11, pp. 366-377

Norris, C.B. 1962. Strength of Orthotropic Materials Subjected to Combined Stress. U.S. Forest Products Lab. Rep. 1816, FPL, Madison, Wisconsin.

Ochoa, O.O.; Reddy, J.N. 1992. Finite Element Analysis of Composite Laminates. Kluwer Academic Publishers, Dordrecht; Boston

Oudjehane, A.; Lam, F.; Avramidis, S. 1998. Forming and Pressing Processes of Random and Oriented Wood Composite Mats. Composites Part B: Engrg., 29B, pp. 211-215

Owen, D.R.J.; Hinton, E. 1980. Finite Elements in Plasticity. McGraw-Hill, New York

Perkins, R.W. 1967. Fundamental Concepts Concerning the Mechanics of Wood Deformation: Strength and Plastic Behaviour. Forest Prod. J., 17(4), pp. 57-68

Pagano, N.J.; Pipes, R.B. 1971. The influence of Stacking Sequence on Laminate Strength. J. Comp. Mat., 5, pp. 50-57

Pipes, R.B.; Cole, B.W. 1973. On the Off-Axis Strength Test for Anisotropic Materials. J. Comp. Mat., 7, pp. 246-256

Rowlands, R.E. 1985. Strength (Failure) Theories and Their Experimental Correlation. Handbook of Composites, Vol.3 - Failure Mechanics of Composites, pp. 71-125

Sandhu, R.S. 1972. A Survey of Failure Theories of Isotropic and Anisotropic Materials. U.S.Air Force Tech. Report No. AFFDL-TR-72-71, Wright Patterson AFB, OH

Shaler, S.M.; Blankenhorn P.R., 1990. Composite Model Prediction of Elastic Moduli for Flakeboard. Wood and Fiber Science, 22(3), 1990, pp.246-261

Sharp, D.J.; Suddarth, S.K. 1991. Volumetric effects in structural composite lumber. In: Proceedings, Int. Timber Eng. Conf., London, England. 3, pp. 427-433

Sharp D.J. 1996. Manufactured Structural Composite Lumber. Inter. COST 508 Wood Mech. Conf., Stuttgart Germany, pp. 273-283

Shih, C.F.; Lee, D. 1978. Further Developments in Anisotropic Plasticity. Trans. ASME, J. Eng. Mat. and Tech., 100, pp. 294-302

Siau, J.F. 1984. Transport Processes in Wood. Springer Verlag, Berlin

Sokolnikoff, I.S. 1956. Mathematical Theory of Elasticity, McGraw-Hill, New York

Suhling, J.C.; Rowlands, R.E.; Johnson, M.W.; Gunderson, D.E. 1984. Tensorial Strength Analysis of Paperboard. Experimental Mech., 25(1), pp. 75-84

Timoshenko, 1972. Mechanics of Materials, New York: Van Nostrand Reinhold, pp. 306-309

Triche M. H.; Hunt M.O. 1993. Modeling of Parallel-Aligned Wood Strand Composites. Forest Products J., 43(11/12), pp. 33-44

Tsai, S. W.; Wu, E. M. 1971. A General Theory of Strength for Anisotropic Materials. J. of Comp. Mat., 5, pp. 58-80.

van der Put, T.A.C.M. 1982. A General Failure Criterion for Wood. IUFRO Timber Engineering Group Meeting, Paper 23 [Sweden], IUFRO, Vienna

Vaziri, R.; Olson, M.D.; Anderson, D.L. 1991. A Plasticity-Based Constitutive Model for Fibre-Reinforced Composite Laminates. J. of Comp. Mat., 25, pp.512-535

Wang Y. T.; Lam, F. 1998. Computational Modeling of Material Failure for Parallel-Aligned Strand Based Wood Composites. Computational Material Science 11, pp.157-165

Weibull, W. 1939. A Statistical Theory of the Strength of Materials. Proc., Royal Swedish Inst., No. 151, Stockholm, Sweden

Whang, B. 1969 Elasto-Plastic Orthotropic Plates and Shells. Proc. Symposium on Application of Finite Element Method in Civil Engineering, Vanderbilt University Tennessee, pp. 481-515

Whitney, J.M. and Browning, C.E. 1972. Free-Edge Delamination of Tensile Coupons. J. of Comp. Mat., 6, pp.300-303.

Wu, E. M. 1974. Phenomenological Anisotropic Failure Criterion Mechanics of Composite Materials (ed. G.P. Sendeckyj) New York: Academic Press

Yang, T. Y. 1986. Finite Element Structural Analysis. Prentice-Hall, Inc., Engle wood Cliffs, New Jersey

Zweben, C. 1994. Size Effect in Composite Materials and Structures: Basic Concepts and D e s i g n Considerations. NASA Conference Publication (Vol/Iss:3271) pp. 197-217

Zienkiewicz, O.C. and Taylor, R.L. 1971. The Finite Element Method: Volume 1, Basic Formulation and Linear Problems. McGraw-Hill Book Company, London

"To determine the nature of the subsequent yield surfaces is one of the major problems in the work-hardening theory of plasticity" (Chen and Han, 1988). The manner in which the yield surfaces evolve beyond initial yield is dictated by the hardening rule. Many rules have been proposed, the most widely known being the simple extremes, isotropic and kinematic hardening as described in Chapter 3. In this study, we investigate an anisotropic rule whereby the subsequent yield surfaces satisfy

$$f \equiv \overline{\sigma}^2(\sigma_i, \alpha_i, M_{ij}(\chi)) - k^2(\chi) = 0 \tag{A.1}$$

such that k varies proportionally with effective plastic strain per Equation 3.45 and the nonlinear parameters of M_{ij} vary non-proportionally described as follows:

Only two strength variables evolve plastically in our case, X_c and Y_c. The other strength variables are assumed to fail in an abrupt brittle manner. Following a procedure initially proposed by Whang (1969), X_c and Y_c are updated upon detection of yielding and for each subsequent iteration. The updated values are calculated by equating the plastic work done (W^P) during plastic deformation in a uniaxial test to that produced by the effective stress and effective plastic strain for a multi-axial state of stress.

Referencing Figure A.1, the plastic work done in a uniaxial test which is approximated by a bilinear stress - plastic strain curve is

$$W^P = \frac{1}{2E_p}(\Gamma_i^2 - \Gamma_{o1}^2) \tag{A.2}$$

where E_p is the plastic modulus and Γ_i and Γ_{oi} represent the initial and subsequent yield values for either X_c or Y_c.

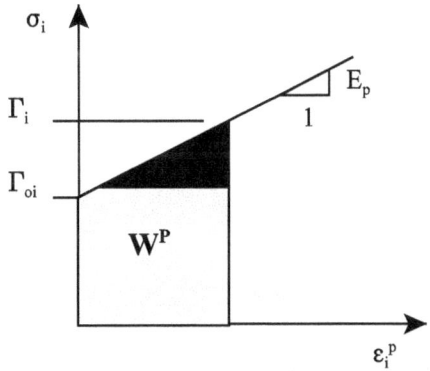

Figure A.1 - Bilinear Stress - Plastic Strain Curve

In the same way, the work done by the effective yield stress is

$$W^P = \frac{1}{2H'}(k^2 - k_o^2) \tag{A.3}$$

It is noted that the effective stress - effective strain relationship is obtained from either one of the uniaxial stress-strain curves (ie. parallel-to-grain or perpendicular-to-grain). The analysis is independent of choice made.

Equating Equation A.2 to A.3, we have

$$\frac{1}{2H'}\left(k^2 - k_o^2\right) = \frac{1}{2E_{pi}}\left(\Gamma_i^2 - \Gamma_{oi}^2\right) \tag{A.4}$$

Rearranging, the updated values for X_c or Y_c is found to be

$$\Gamma_i^2 = \frac{E_{pi}}{H'}\left(k^2 - k_o^2\right) + \Gamma_{oi}^2 \tag{A.5}$$

or specifically,

$$X_{cr}^2 = \frac{E_{p1}}{H'}\left(k_r^2 - k_o^2\right) + X_c^2 \tag{A.6a}$$

$$Y_{cr}^2 = \frac{E_{p2}}{H'}\left(k_r^2 - k_o^2\right) + Y_c^2 \tag{A.6b}$$

A preliminary study was carried out to establish a computer-time efficient yet numerically accurate finite element mesh for the uniaxial analyses. Three mesh sizes using simple rectangular elements were investigated. In one case, a rather coarse mesh was used dividing the x and y dimensions into thirds (3 x 3 grid). The second was refined to a 4 x 4 grid and the third was refined further to a 5x5 grid. All three are illustrated in Figure B.1.

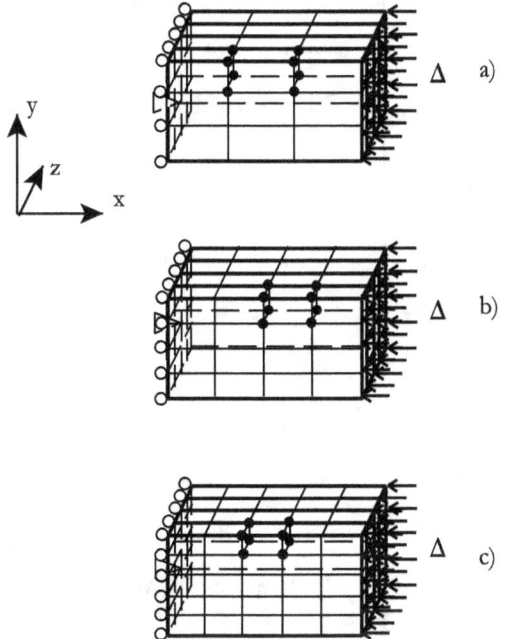

Figure B.1 - **Various Mesh Grades Investigated for Uniaxial Analyses: a) 3 x 3 grid, b) 4 x 4 grid, c) 5 x 5 grid**

A $[\pm 30]_s$, angle-ply laminate subjected to compressive displacement was modeled using the 3 dimensional program. To ensure comparability between meshes, a deterministic analysis was performed. The same strength and stiffness variables, outlined in Table B.1, were used for each case.

Table B.1 - Input Properties for Parametric Study

Property		Mean	St. Dev.
Parallel-to-grain Tension	Elastic Modulus (E_{Xt}) (MPa)	15463	0
	Strength (X_t) (MPa)	72.8	0
Perpendicular-to-grain Tension	Elastic Modulus (E_{Yt}) (MPa)	91.2	0
	Strength (Y_t) (MPa)	6.5	0
Parallel-to-grain Compression	Elastic Modulus (E_{Xc}) (MPa)	10090	0
	Yield Strength (X_c) (MPa)	67.3	0
	Tangent Modulus (E_{Xc}') (MPa)	1926	0
	Ultimate Strength (X_c^u) (MPa)	76.5	0
Perpendicular-to-grain Compression	Elastic Modulus (E_{Yc}) (MPa)	490	0
	Yield Strength (Y_c) (MPa)	15.4	0
	Tangent Modulus (E_{Yc}') (MPa)	110	0
	Ultimate Strength (Y_c^u) (MPa)	18.2	0
Interaction Parameter	F_{12}	1.067×10^{-03}	0
In plane Shear	Elastic Modulus (G) (MPa)	392.2	0
	Strength (S) (MPa)	11.4	0
Poisson's Ratio	ν_{12}	0.32	-

The results of the three analyses are given graphically in Figure B.2. The ultimate load for each case varies slightly with progressively lower predictions for the finer grids (26.8 MPa, 26.3 MPa and 25.5 MPa for the coarse, average and fine mesh, respectively). This was a result of stress concentrations at the boundaries. The smaller elements produced higher stresses for equal displacement and thus failed earlier. Considering the results were relatively consistent, it was decided that the best element for both efficiency and accuracy would be the 4 x 4 grid.

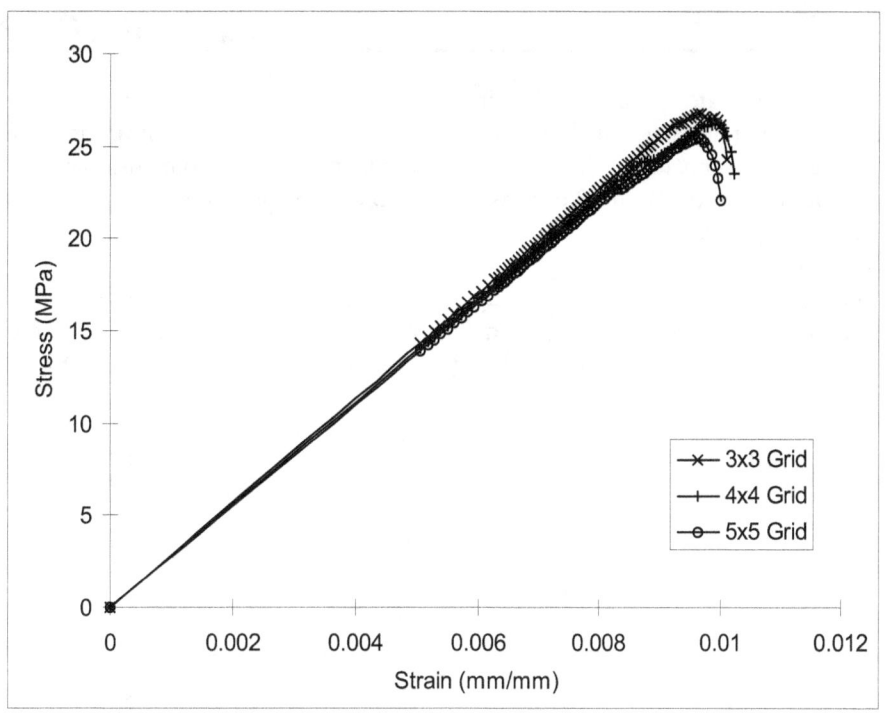

Figure B.2 - Stress - Strain Diagrams of 3 Investigated Grids

A stochastic, materially nonlinear finite element based FORTRAN 77 program has been developed for this study. The program, entitled COMAP (COMposite Analysis Program), has extended capacity to perform Monte Carlo simulations. The program was adapted from a 2-dimensional nonlinear program by Owen and Hinton (1980). Both a 2-dimensional and 3-dimensional version of the program have been developed and investigated.

This Appendix describes the most important aspects of the program. A flowchart highlighting the essential steps in the program is provided in Figure C.1. Sample input and output files for the 3 dimensional program have been included and described for completeness.

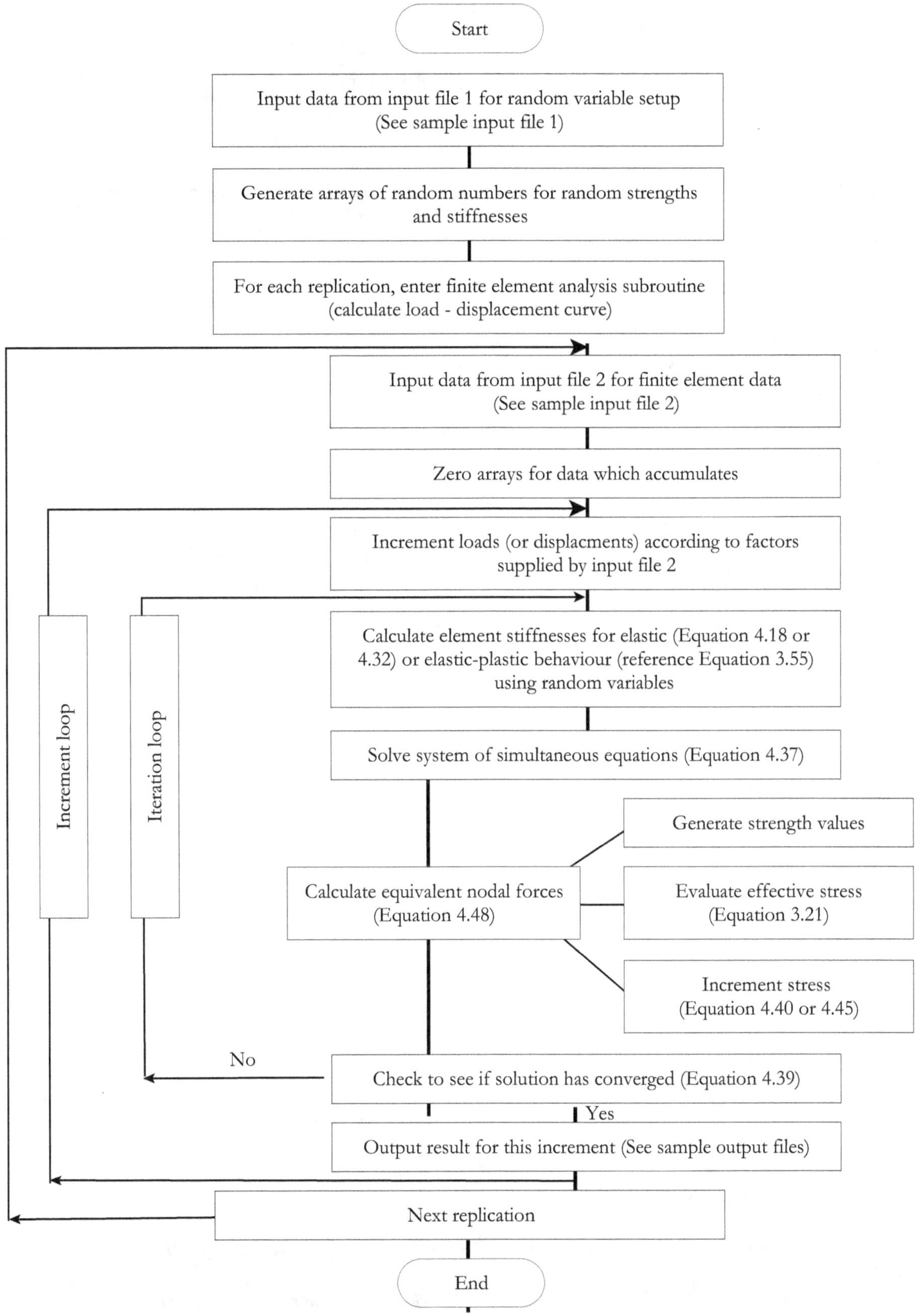

Figure C.1 - Program Flowchart (Adapted from Owen and Hinton, 1980)

Line #

1	1.00E+00	*Line 1-32:*	*Values of lower triangle correlation matrix for compression properties (Tables 5.5 and 5.7)*
2	0.00E+00		
3	0.00E+00		
4	0.00E+00		
5	4.77E-01		
6	8.79E-01		
7	0.00E+00		
8	0.00E+00		
9	2.00E-01		
10	-3.52E-01		
11	9.14E-01		
12	0.00E+00		
13	5.54E-01		
14	6.54E-01		
15	2.84E-01		
16	4.30E-01		
17	1.00E+00		
18	0.00E+00		
19	0.00E+00		
20	0.00E+00		
21	1.33E-01		
22	9.91E-01		
23	0.00E+00		
24	0.00E+00		
25	1.05E-01		
26	-4.33E-01		
27	8.95E-01		
28	0.00E+00		
29	3.69E-01		
30	7.66E-01		
31	4.04E-01		
32	3.38E-01		

33	2	*Line 33:*	*Number of input files for finite element analysis*

34	3D_3pnt15.txt	*Lines 34-35:*	*Name of input files for finite element analysis*
35	3D_3pnt30.txt		

36	1408	4	*Line 36:*	*Number of Gauss points per layer, number of layers*

37	50	*Line 37:*	*Number of replications per random seed pair*

38	10	*Line 38:*	*Number of random seeds*

39	111	222	*Lines 39-48: Random seeds*
40	222	333	
41	333	444	
42	444	555	
43	555	666	
44	666	777	
45	777	888	
46	888	999	
47	123	456	
48	456	789	

INPUT FILE 2 (SAMPLE)

Line #

Line #										
1	15 angle ply									
2	345	176	15	8	4	2	1	300	6	17
3	1	1	1	4	5	2	16	19	20	18
4	2	1	2	5	6	3	17	20	21	20
5	3	2	4	7	8	5	19	22	23	20
6	4	2	5	8	9	6	20	23	24	21
7	5	3	7	10	11	8	22	25	26	23
178	175	4	325	328	329	326	340	343	344	341
179	176	4	326	329	330	327	341	344	345	342
180	1	0	0	0						
181	2	0	9.5	0						
182	3	0	19	0						
183	4	0	0	2.75						
184	5	0	9.5	2.75						
185	6	0	19	2.75						
523	344	228	9.5	11						
524	345	228	19	11						
525	16	110	0	0	0					
526	19	110	0	0	0					
527	22	111	0	0	0					
528	25	110	0	0	0					
529	28	110	0	0	0					
530	168	10	0	-0.3	0					
531	171	10	0	-0.3	0					
532	174	10	0	-0.3	0					
533	177	10	0	-0.3	0					
534	180	10	0	-0.3	0					

```
535   316        10         0      0       0    0
536   319        10         0      0       0    0
537   322        11         0      0       0    0
538   325        10         0      0       0    0
539   328        10         0      0       0    0
540   1
541   10090      0.3        0.0    15.0    67.3    1926.0   134.4   490.0   392.2   15.4   11.4   5.8   110.0   1.07E-03   76.5   18.2
542   35.86      1.1        13.0   5.4     1.8     1.7      1930.0  639.0   22.3    0.0    2.5
543   9.38E-05   85.0       15463.0  91.2  4718.0  22.3     0.0     1.0     0.8
544   2
545   10090      0.3        0.0    -15.0   67.3    1926.0   134.4   490.0   392.2   15.4   11.4   5.8   110.0   1.07E-03   76.5   18.2
546   35.86      1.1        13.0   5.4     1.8     1.7      1930.0  639.0   22.3    0.0    2.5
547   9.38E-05   85.0       15463.0  91.2  4718.0  22.3     0.0     1.0     0.8
548   3
549   10090      0.3        0.0    -15.0   67.3    1926.0   134.4   490.0   392.2   15.4   11.4   5.8   110.0   1.07E-03   76.5   18.2
550   35.86      1.1        13.0   5.4     1.8     1.7      1930.0  639.0   22.3    0.0    2.5
551   9.38E-05   85.0       15463.0  91.2  4718.0  22.3     0.0     1.0     0.8
552   4
553   10090      0.3        0.0    15.0    67.3    1926.0   134.4   490.0   392.2   15.4   11.4   5.8   110.0   1.07E-03   76.5   18.2
554   35.86      1.1        13.0   5.4     1.8     1.7      1930.0  639.0   22.3    0.0    2.5
555   9.38E-05   85.0       15463.0  91.2  4718.0  22.3     0.0     1.0     0.8
556   bending
557   0          0          0
558   0.01       1          30     0       3
559   3.8        1          30     0       3
560   0.08       1          60     0       3
561   0.08       1          60     0       3
562   0.08       1          60     0       3
563   0.08       1          60     0       3
856   0.08       1          60     0       3
857   0.08       1          60     0       3
```

Input file 1 Explanation:

Line						
Line 1:	Title					
Line 2:	# points	# elements	# boundary conditions	# nodes / element	# materials	Order of integration
	Stiffness parameter	# increments	# stress components			
Line 3-179:						
Line 180-524:	Element #	Material #	Element node numbers	z coordinate		
		x coordinate	y coordinate			
Line 525-539:	Node #	Restraint code	Prescribed x displacement	y displacement	z displacement	
Line 540:	Material #					
Line 541:	EX_c (mean)	$n12$	thickness	angle	X_c (mean)	EX_c' (mean)
	X_t (mean)	EY_c (mean)	G (mean)	Y_c (mean)	S (mean)	Y_t (mean)
	EY_c' (mean)	$F12$ (mean)	XC_u (mean)	YC_u (mean)		
			EX_t (mean)	EY_t (mean)		
Line 542:	X_t (st.dev.)	Y_t (st.dev.)	X_c (st.dev.)	X_{cu} (st.dev.)	Y_c (st.dev.)	Y_{cu} (st.dev.)
	Ex_c (st.dev.)	Ex_c' (st.dev.)	Ey_c (st.dev.)	Ey_c' (st.dev.)	S (st.dev.)	EY_t (st.dev.)
	$F12$ (st.dev.)	G (st.dev.)			EX_t (st.dev.)	
Line 543:	Failure mode parameter	Extrapolation parameter				
Line 544:	Material #					
Line 545-547:	Material statistical properties as for lines 541-543					
Line 548:	Material #					
Line 549-551:	Material statistical properties as for lines 541-543					
Line 552:	Material #					
Line 553-555:	Material statistical properties as for lines 541-543					
Line 556:	Load case title					
Line 557:	Load control parameters	Convergence tolerance	Maximum # of iterations	Output parameters		
Line 558-857:	Increment factor					

OUTPUT FILE 1 (SAMPLE)

The output file described herein is a portion of the finite element results for one replication. For normal use, this information is not output due to the extreme size of the file for many replications.

```
15 angle ply

        NUMBER OF NODES............NPOIN=   345
        NUMBER OF ELEMENTS.........NELEM=   176
        NUMBER OF RESTRAINED POINTS...NVFIX=   15
        NUMBER OF NODES PR. ELEMENT...NNODE=    8
        NUMBER OF DIFFERENT MATERIALS..NMATS=    4
        ORDER OF INTEGRATION FORMULA...NGAUS=    2
        TOTAL DEGREES OF FREEDOM......NEVAB=   24
        NONLINEAR SOLUTION PARAMETER...NALGO=    1
        NUMBER OF LOADING INCREMENTS...NINCS=  300
        NUMBER OF STRESS COMPONENTS...NSTRE=    6


ELEMENT   PROPERTY      NODE NUMBERS

   1         1        1    4    5    2   16   19   20   17
   2         1        2    5    6    3   17   20   21   18
   3         2        4    7    8    5   19   22   23   20
   4         2        5    8    9    6   20   23   24   21
   5         3        7   10   11    8   22   25   26   23

 175         4      325  328  329  326  340  343  344  341
 176         4      326  329  330  327  341  344  345  342
1

NODE      X          Y          Z
   1   .00000    .00000    .00000
   2   .00000   9.5000    .00000

 344  228.00    9.5000   11.000
 345  228.00   19.000    11.000

NODE   CODE         FIXED VALUES
  16   110    .000    .000    .000
  19   110    .000    .000    .000
```

22	111	.000	.000	.000
25	110	.000	.000	.000
28	110	.000	.000	.000
168	10	.000	-.300	.000
171	10	.000	-.300	.000
174	10	.000	-.300	.000
177	10	.000	-.300	.000
180	10	.000	-.300	.000
316	10	.000	.000	.000
319	10	.000	.000	.000
322	11	.000	.000	.000
325	10	.000	.000	.000
328	10	.000	.000	.000

bending

INCREMENT NUMBER   1

DISPLACEMENTS

NODE	X-DISP.	Y-DISP.	Z-DISP.
1	.496900E-05	.708890E-03	-.627791E-05
2	.383505E-03	.713694E-03	.259566E-05
3	.785140E-03	.703390E-03	.107422E-04
4	-.741297E-05	.729206E-03	-.667442E-05
344	.397716E-03	.701700E-03	.600917E-04
345	.821824E-05	.673114E-03	.446213E-04

REACTIONS

NODE	X-REAC.	Y-REAC.	Z-REAC
16	.803979E-01	.498102E-01	.000000
19	-.313343E-01	.121242	.000000
22	-.248297	.128356	-.353298E-02
25	.249075E-01	.155893	.000000
28	.174325	.868924E-01	.000000
168	.000000	-.582413E-01	.000000
171	.000000	-.353728	.000000
174	.000000	-.121165	.000000
177	.000000	-.481618	.000000
180	.000000	-.696373E-01	.000000
316	.000000	.413438E-01	.000000

```
319    .000000    .165812      .000000
322    .000000    .903560E-01  .353298E-02
325    .000000    .198999      .000000
328    .000000    .456834E-01  .000000
```

1 G.P.	XX-STRESS	YY-STRESS	ZZ-STRESS	XY-STRESS	XZ-STRESS	YZ-STRESS	E.P.S.	11-STRESS	22-STRESS	33-STRESS	12-STRESS	23-STRESS	13-STRESS

ELEMENT NO.... 1

	XX-STRESS	YY-STRESS	ZZ-STRESS	XY-STRESS	XZ-STRESS	YZ-STRESS	E.P.S.	11-STRESS	22-STRESS	33-STRESS	12-STRESS	23-STRESS	13-STRESS
1	.2275706E-02	.7463246E-04	-.1159301E-04	.8790223E-03	-.3205452E-03	.1059334E-03	.000000E+00	.3101910E-02	-.1790854E-03	.1159301E-04	.6786563E-04	-.2822054E-03	.1852870E-03
2	.4390741E-02	.4941004E-04	-.9729713E-05	.1301884E-02	-.3076611E-03	.1091332E-03	.000000E+00	.5781433E-02	-.2367296E-03	-.9729713E-05	-.2340066E-03	-.2689320E-03	.1850431E-03
3	-.8054703E-03	-.4482476E-03	-.2107101E-04	-.6200181E-03	-.9590830E-04	.8952915E-04	.000000E+00	-.1280604E-02	-.1757414E-03	.2107101E-04	-.3968988E-03	-.1158122E-03	.6165563E-03
4	.3709994E-03	-.5221766E-03	-.2421548E-04	-.3971388E-03	.1080373E-03	.9254825E-04	.000000E+00	.1996769E-03	-.2575238E-03	.2421548E-04	.5905588E-03	.1283093E-03	.6143263E-04
5	.5481477E-03	-.8730056E-03	-.1490469E-04	-.5242009E-04	.4650232E-04	.5845519E-04	.000000E+00	.5553957E-03	-.7423592E-03	.1490469E-04	.4351591E-03	.6004711E-04	.4442769E-04
6	-.2207325E-02	-.9267705E-03	-.1289991E-04	.1983061E-03	.5902511E-04	.6139647E-04	.000000E+00	-.2614621E-02	-.7787820E-03	.1289991E-04	-.7506068E-04	-.7290446E-04	.4402761E-04
7	-.2533730E-02	-.1396089E-02	-.2742078E-04	-.1551604E-02	.2135091E-03	.8235365E-05	.000000E+00	-.3828024E-02	-.7391920E-03	.2742078E-04	-.8999693E-04	-.2083701E-03	-.4728799E-04
8	-.1813119E-02	-.1498561E-02	-.3042377E-04	-.1150861E-02	.2252767E-03	.1101416E-03	.000000E+00	-.2968040E-02	-.7997557E-03	.3042377E-04	-.1107115E-02	.2204513E-03	-.4766705E-04

ELEMENT NO.... 2

	XX-STRESS	YY-STRESS	ZZ-STRESS	XY-STRESS	XZ-STRESS	YZ-STRESS	E.P.S.	11-STRESS	22-STRESS	33-STRESS	12-STRESS	23-STRESS	13-STRESS
1	.8423729E-03	-.1516655E-05	-.7758530E-05	.8986361E-04	.2158421E-04	.6379883E-04	.000000E+00	.1018432E-02	-.1158140E-03	.7758530E-05	-.2236632E-03	.2249997E-03	.5760448E-05
2	.6012482E-03	-.2699988E-03	-.1097223E-04	-.1218933E-03	.2263467E-03	.6653824E-04	.000000E+00	.6230598E-03	-.1405577E-03	.1097223E-04	-.3611876E-03	.2358555E-03	.5688162E-05
3	-.1467273E-02	-.3110719E-03	-.1555102E-03	-.4945429E-03	.1229004E-03	.6933720E-05	.000000E+00	-.1981481E-02	-.1659771E-03	.1555102E-03	-.4695809E-04	-.1205073E-03	-.2511151E-04
4	-.1945970E-02	-.4187518E-03	-.2034229E-03	-.7035929E-03	.1209942E-03	.8642147E-05	.000000E+00	-.2652266E-02	-.2020523E-03	.2034229E-03	-.1051407E-03	.1269322E-03	-.2506436E-04
5	-.2971093E-03	-.3131276E-03	-.2695503E-04	-.3723692E-03	.2358340E-03	-.4923772E-05	.000000E+00	-.5541023E-03	-.1307867E-03	.2695503E-04	-.3078002E-03	.2265238E-03	-.6579433E-04
6	-.6382697E-03	-.3844214E-03	.2723442E-04	-.6043630E-03	.2442757E-03	-.3659558E-05	.000000E+00	-.1073257E-02	-.1100004E-03	-.2723442E-04	-.4197902E-03	.2350051E-03	-.6675807E-04
7	-.2610761E-02	-.4736956E-03	-.1819465E-05	-.9575968E-03	.7656328E-04	-.2881976E-04	.000000E+00	-.3559132E-02	-.1820490E-03	.1819465E-05	-.1308432E-03	.6649535E-04	-.4765379E-04
8	-.3189493E-02	-.5343361E-03	.5212825E-06	-.1186884E-02	.8069418E-04	-.2858699E-04	.000000E+00	-.4353689E-02	-.1725042E-03	-.5212825E-06	-.1614913E-03	.7054574E-04	-.4849810E-04

...

ELEMENT NO....176

	XX-STRESS	YY-STRESS	ZZ-STRESS	XY-STRESS	XZ-STRESS	YZ-STRESS	E.P.S.	11-STRESS	22-STRESS	33-STRESS	12-STRESS	23-STRESS	13-STRESS
1	.8392611E-02	.5953990E-03	.7938079E-04	.2510253E-02	-.5737584E-04	.9226558E-04	.000000E+00	.1109528E-01	.4016018E-05	-.7938079E-04	-.3031812E-03	-.3154071E-04	.1039717E-03
2	.6289744E-02	.3560229E-02	.5867784E-04	.1995896E-02	-.6486502E-04	.8677400E-04	.000000E+00	.8366491E-02	-.1384489E-03	-.5867784E-04	-.1505022E-03	-.4019604E-04	.1006056E-03
3	.9819044E-03	.8533448E-05	-.1381832E-04	.4449075E-03	-.4033815E-04	-.2485068E-04	.000000E+00	.1369620E-02	-.1321701E-03	.1381832E-04	.8020539E-04	-.4539549E-04	-.1353633E-04
4	.5966297E-03	-.9781154E-04	-.2531666E-04	.4042978E-03	-.3528487E-04	-.2734126E-04	.000000E+00	.8922965E-03	-.2433875E-03	.2531666E-04	.1389991E-03	-.4115901E-03	-.1727723E-04
5	.4575488E-02	.2732149E-04	-.5120341E-04	.1117447E-02	-.4348000E-03	.5425179E-04	.000000E+00	.5903468E-02	-.1496283E-03	.5120341E-04	-.4570615E-04	-.4059431E-03	.1649377E-03
6	.3359150E-02	-.8452327E-04	.3696267E-04	.9246900E-03	-.4362872E-03	.5305366E-04	.000000E+00	.4379248E-02	-.2595788E-03	-.3696267E-04	-.2713738E-03	-.4076898E-03	.1641653E-03
7	-.2815471E-02	-.5558191E-03	-.4973661E-06	-.9432724E-03	-.1374610E-03	.3196032E-04	.000000E+00	-.3796566E-02	-.2829916E-03	.4973661E-06	-.7491684E-04	-.1245051E-03	.6644882E-04
8	-.2314217E-02	-.5346328E-03	-.5533505E-05	-.6622832E-03	-.1264057E-03	.3376318E-04	.000000E+00	-.3069325E-02	-.3616990E-03	.5533505E-05	-.1688558E-04	-.1133600E-03	.6532893E-04

INCREMENT NUMBER   2

...

OUTPUT FILE 2 (SAMPLE)

Following is a typical output file providing the load - displacement curve for many replications ...

```
DISPLACEMENT    LOAD


  -.300000E-02    .130979E+01
  -.114300E+01    .413121E+03
  -.116700E+01    .421791E+03
  -.119100E+01    .430460E+03
        ⋮
  -.282300E+01    .963515E+03
  -.284700E+01    .967138E+03
  -.287100E+01    .969484E+03
  -.289500E+01    .969407E+03
  -.291900E+01    .970101E+03
  -.294300E+01    .970393E+03
  -.296700E+01    .969730E+03
  -.299100E+01    .965100E+03
  -.301500E+01    .958432E+03
  -.303900E+01    .949766E+03
  -.306300E+01    .935536E+03
  -.308700E+01    .917108E+03
  -.311100E+01    .899986E+03
  -.313500E+01    .884894E+03
  -.315900E+01    .870838E+03

DISPLACEMENT    LOAD


  -.300000E-02    .115447E+01
  -.114300E+01    .416110E+03
  -.116700E+01    .424846E+03
  -.119100E+01    .433582E+03
        ⋮
  -.248700E+01    .853558E+03
  -.251100E+01    .855299E+03
  -.253500E+01    .852577E+03
  -.255900E+01    .846192E+03
  -.258300E+01    .838516E+03
  -.260700E+01    .831685E+03
  -.263100E+01    .824157E+03
  -.265500E+01    .816414E+03
  -.267900E+01    .806546E+03
  -.270300E+01    .794212E+03
  -.272700E+01    .780463E+03
  -.275100E+01    .766277E+03
        ⋮
```